Springer Proceedings in Mathematics & Statistics

Volume 135

More information about this series at http://www.springer.com/series/10533

Springer Proceedings in Mathematics & Statistics

This book series features volumes composed of select contributions from workshops and conferences in all areas of current research in mathematics and statistics, including OR and optimization. In addition to an overall evaluation of the interest, scientific quality, and timeliness of each proposal at the hands of the publisher, individual contributions are all refereed to the high quality standards of leading journals in the field. Thus, this series provides the research community with well-edited, authoritative reports on developments in the most exciting areas of mathematical and statistical research today.

Jaime A. Londoño • José Garrido
Daniel Hernández-Hernández
Editors

Actuarial Sciences and Quantitative Finance

ICASQF, Bogotá, Colombia, June 2014

Springer

Editors
Jaime A. Londoño
Department of Mathematics & Statistics
National University of Colombia
Manizales, Colombia

José Garrido
Department of Mathematics and Statistics
Concordia University
Montréal, Québec, Canada

Daniel Hernández-Hernández
Mathematics Research Center (CIMAT)
Guanajuato, Mexico

ISSN 2194-1009 ISSN 2194-1017 (electronic)
Springer Proceedings in Mathematics & Statistics
ISBN 978-3-319-18238-4 ISBN 978-3-319-18239-1 (eBook)
DOI 10.1007/978-3-319-18239-1

Library of Congress Control Number: 2015942912

Springer Cham Heidelberg New York Dordrecht London

Printed on acid-free paper

Springer International Publishing AG Switzerland is part of Springer Science+Business Media (www.springer.com)

Preface

The chapters in this volume of the Springer Proceedings in Mathematics & Statistics titled "Advances in Actuarial Sciences and Quantitative Finance" from the proceedings of the First International Congress on Actuarial Science and Quantitative Finance, which took place in Bogotá from June 17 to 20, 2014. The conference was organized by the Department of Mathematics of the Universidad Nacional de Colombia with the support of other academic institutions that include Universidad Antonio Nariño, Universidad Externado de Colombia and Universidad del Rosario. Additional sponsors include COLCIENCIAS, ICETEX, Fasecolda and Asociación Colombiana de Actuarios. The conference took place in the Centro Cultural Gabriel Garcia Marquez and Biblioteca Luis Ángel Arango. The scientific committee consisted of Prof. Samuel Cox (University of Manitoba, Canada), Prof. José Garrido (Concordia University, Canada), Prof. Daniel Hernández-Hernández (CIMAT, Mexico), Prof. Jaime Londoño (Universidad Nacional, Colombia), Prof. Javier Peña (Carnegie Mellon University, USA), Prof. Jean-Luc Prigent (Université de Cergy-Pontoise, France) and Prof. Armando Zarruk (Universidad Nacional, Colombia).

This was the first edition of a series of events to be organized every other year, whose objective is to become a reference in Actuarial Science and Quantitative Finance in Colombia, the Andean region (Peru, Colombia, Venezuela, Ecuador, and Bolivia) and the Caribbean, with participation from researchers, students and practitioners from different parts of the world. This first edition contributed to enhancing the relations between the academic and industrial actuarial and financial communities of Colombia and the Andean Region with the corresponding communities in North America, Europe and other parts of the world.

The emphasis of the event was equally distributed between Actuarial Science and Quantitative Finance and covered a variety of topics such as Statistical Techniques in Finance and Actuarial Science, Portfolio Management, Derivative Valuation, Risk Theory and Life and Pension Insurance Mathematics, Non-Life Insurance Mathematics, and Economics of Insurance.

The event consisted of plenary sessions with invited speakers in the areas of Actuarial Science and Quantitative Finance, oral sessions of contributed talks on these topics, as well as poster sessions and short courses taught by some of the invited speakers. The list of invited speakers reflects the broad variety of topics: Monique Jeanblanc (Arbitrages in a Progressive Enlargement of Filtrations), Stephane Loisel (From Ruin Theory to Longevity Risk), Hansjoerg Albrecher (Insurance Risk, Bankruptcy and the Cost of Capital), Steve Haberman (Longevity Risk and Longevity Trends), Ajay Subramanian (Dynamic Prudential Regulation), Fabio Mercurio (Derivative Pricing Under Collateralization and Differential Rates), David Ingram (Culture and its Impact on Risk Management of an Insurance Enterprise), Richard Davis (Noncausal Vector AR Processes with Application to Economic Time Series), Edward (Jed) Frees (Insurance Company Operations and Dependence Modeling) and Carlos Vazquez Cendón (Speed up of Calibration and Pricing with SABR Models: From Equities to Interest Rates Derivatives). Topics for short courses included: Heavy-tailed Time Series: Theory and Applications (Richard Davis), Regression Modelling with Actuarial Applications (Edward W. Frees), Modelling Credit Risk Events (Monique Jeanblanc) and Enterprise Risk Management for Insurance Companies (Stephane Loisel).

Additionally, contributions from researchers and students were presented in oral and poster sessions. There were 37 oral presentations from 56 submitted contributions, and three poster presentations out of 6 submitted poster contributions. Speakers came from a diverse list of countries that included Argentina, Brazil, Canada, Colombia, Costa Rica, France, Ireland, Mexico, Netherlands, Taiwan, USA, Spain, Switzerland, UK and Venezuela. The number of contributions along with a total number of 249 registered participants outnumbered our expectations for the first edition of this event.

The Congress gave an emphasis on the relation between industry and academia providing a day to address problems arising from the financial and insurance industries. As a matter of fact, topics of talks, and in some cases the speakers themselves came from those industries. The space provided to practitioners a platform to present and discuss with academics and students different approaches to addressing problems arising from the industries in the region.

The current proceedings are based on invitations to selected oral contributions and selected contributions presented by the invited speakers. All contributions were subject to a review process. The spectrum of the five papers published here reflects the diverse nature of the presentations. We are grateful to the authors who submitted papers for this volume as well as to the reviewers for their expertise.

Special thanks go to the members of the Organizing Committee, which included Prof. Carlos Alberto Castro (Universidad del Rosario, Bogotá), Prof. Johanna Garzón (Universidad Nacional de Colombia, Bogotá), Prof. Jaime

Londoño (Universidad Nacional de Colombia, Bogotá), Prof. Victor Hugo Prieto (Universidad Antonio Nariño, Bogotá), Prof. Alejandra Sánchez (Universidad Nacional de Colombia, Bogotá) and Prof. Javier Sandoval (Universidad Externado de Colombia, Bogotá). Finally, we would like to thank all the conference participants who made this event a great success.

Montréal, QC, Canada	José Garrido
Guanajuato, Mexico	Daniel Hernández-Hernández
Manizales, Colombia	Jaime A. Londoño

Contents

Contributors

Viswanathan Arunachalam Department of Statistics, National University of Columbia, Bogotá D.C., Colombia

James Beatty KPMG, New York, USA

David Bogataj The European Faculty of Law, Nova Gorica, Slovenia

Marija Bogataj The Mediterranean Institute for Advanced Studies, Sempeter pri Gorici, Slovenia

Rodrigo Cancino Seguros Bolívar, Bogotá D.C., Colombia

Ana María Ferreiro University of A Coruña, A Coruña, Spain

José A. García-Rodríguez University of A Coruña, A Coruña, Spain

Argyn Kuketayev E*Trade Financial, Vienna, USA

José G. López-Salas University of A Coruña, A Coruña, Spain

Diego Ros McDonnell Cartagena Polytechnic University, Cartagena, Spain

Carlos Vázquez University of A Coruña, A Coruña, Spain

Fabio Mercurio Bloomberg L.P., New York, USA

Modeling Electricity Spot Price Dynamics by Using Lévy-Type Cox Processes: An Application to Colombian Market

Viswanathan Arunachalam and Rodrigo Cancino

Abstract In this chapter, we propose the dynamic behavior of the Colombian electricity spot prices by using a Lévy-type Cox process. The mean-reverting process is frequently applied to electricity markets, due to certain characteristics of the electricity such as spikes which are intrinsic to electricity prices. Modeling spikes for a particular electricity market is a great challenge. We propose a two factor model for the Colombian spot price dynamics, where one factor is driven by a Ornstein Uhlenbeck process based on Brownian motion, which deals with normal variation, while other factor is driven by a Lévy-type Cox process and specifies price spikes. We discuss the methodology of calibration and estimation of parameters of electricity spot prices for the Colombian Markets.

Keywords Electricity spot prices • Cox process • Spikes • Colombian market

1 Introduction

In recent years, a significant importance has been given to pricing and risk management of electricity and related financial derivatives. Electricity is an unique commodity with distinctive characteristics such as seasonality, high volatility, and spikes. Hence, modeling the price behavior of electricity price and forecasting the financial derivatives becomes a challenging task. The physical attributes of electricity is different to commodities such as gold and oil, in sense that it is difficult to store, uncertain and inelastic demand.

In Colombia, the electricity market is composed of four sectors: generation, transmission, distribution, and commercialization. The electricity price, without including transport or commercial charges, is formed through the interaction

---

V. Arunachalam (✉)
Department of Statistics, Universidad Nacional de Colombia, Bogotá, DC, Colombia
e-mail: varunachalam@unal.edu.co

R. Cancino
Seguros Bolívar, Bogotá, DC, Colombia
e-mail: rodrigo.cancino@segurosbolivar.com

J.A. Londoño et al. (eds.), *Actuarial Sciences and Quantitative Finance*, Springer Proceedings in Mathematics & Statistics 135, DOI 10.1007/978-3-319-18239-1_1

between the generators, supply side, and the big consumers and distributors, demand side. Consequently, an understanding of the behavior of the electricity price is important to describe, at least briefly, the composition of the generation sector and the market structure and operation at this level. The generation of electricity is concentrated in hydroelectric plants. This implies that the dynamic of the market is closely related with the behavior of the weather. The installed capacity composition is shown in the following table:

Installed capacity [MW]

Hydroelectric plants:	9,231	64 %
Thermoelectric plants:	4,471	31 %
Minor plants and cogenerators:	721	5 %
Total	14,423	100 %

The "place" where the transactions take place is called "El Mercado de Energía Mayorista" (MEM). The short-term market operates through the "Bolsa de Energía" which is the "market pool" where generators place bids to sell electricity for the next day and the administrator of the market assigns which ones will produce, according to the demand for the next day. The important way in which the energy is sold is through bilateral contracts between great consumers and distributors, and the producers, this is the mid-term market. There exists also a long-term market whose purpose is to guarantee the provision of electricity with continuity and quality.

In the present work we will consider only the short-term price of the energy, that is, the price in the "Bolsa de Energía." In the recent literature, it has been shown that due to the liberalization of the market, the behavior of the electricity price presents some features, which we will describe in the following section. But before we do that, it is important to clarify that, although, the prices are set through the interaction of the market, the market is not completely deregulated. In fact, exists a maximum price, given by the "Precio de Escasez," which is the exercise price of a call option sold by generators to guarantee the reliability in the long-term energy supply.

The electricity presents certain features that make it different from the rest of merchandises of the economy. The principal difference is the non-storability of electricity; so after being produced, the electricity has to be consumed immediately. Another particularity is that in the transportation there are losses in the transmissions lines, which, combined with the high costs of transport, makes economically impossible to deliver electricity far away from the production plants. This implies a local consumption. From the demand side, there exist consumption patterns which determines a seasonal behavior, and this can be daily, monthly, yearly, or even longer. Another characteristic of the demand is that it is highly inelastic with respect to the prices, this means that the impact on the demand from changes in prices is small. All this imply that electricity prices present characteristics as: mean reversion behavior, presence of spike associated with on-peak and off-peak electricity prices, and a seasonal dynamics.

The series of the daily spot prices at "Bolsa de Energía" from 1 Jan 2000 to 31 Dec 2013 are presented in Figure 1. Since the price at "Bolsa de Energía" is fixed per hour, we consider the price at 19:00, a peak hour of consumption.

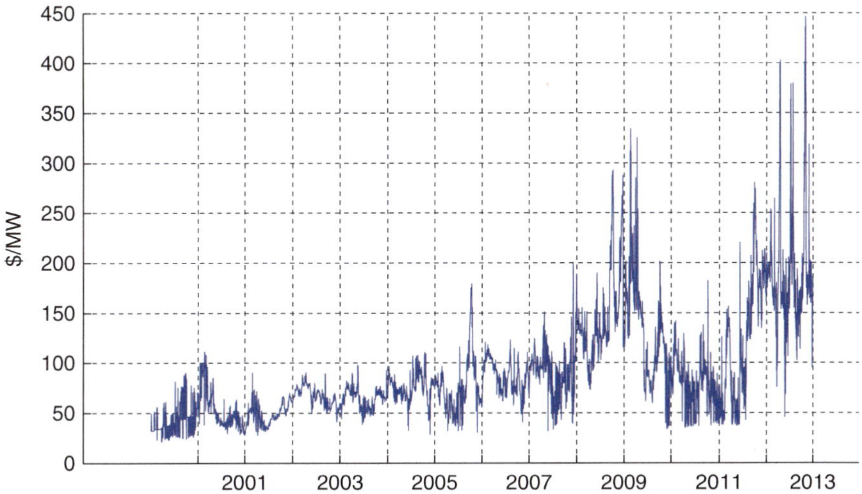

Fig. 1 Price of electricity S_t

The prodigious use of Black-Scholes models driven by Brownian motions has become major source of modeling tool in price dynamics, in particular the mean reversion process or Gaussian Ornstein-Uhlenbeck(OU) process [10, 14] and mean-reverting jump-diffusion model [2, 5, 6, 8]. Recent studies focus on the question "how electricity spot prices are designed in the market?" [4, 7, 12]. More recently, Lévy processes have used modeling for spot price dynamics, in particular the electricity and gas-related spot prices [1, 9, 13]. The unusual characteristics of electricity prices have brought researchers and practitioners to develop specific model of electricity prices for particular market data.

The purpose of this chapter is to study the dynamic behavior of the energy price in the Colombian market through Lévy-type Cox process. We study the effects of both mean-reverting specification of normal variation and spikes for jumps in the spot prices. The difficulty of this study is to detect and filter the spikes in order to estimate the jumps. The method proposed by [3], in which the analysis is made for the German case, inspired us to employ the estimation methodology based on daily spot price data for Colombian Market, which is obtained from the webpage of XM Compañía de Expertos en Mercados S.A ESP.

The chapter is organized as follows: Section 2 is devoted to presenting a stochastic models of electricity spot price evolution based on Lévy-type Cox process. In Section 3, the calibration methodology for separating normal variation and spikes for electricity spot prices to estimate the parameters of the model are presented. Section 4 concludes the chapter.

2 Electricity Spot Price Process

We would like to explain our choice of model that incorporates the best aspects of the Colombian Market data described in Section 1 and address the non-Gaussian jumps based Lévy-type Cox process. The idea is to divide the price dynamics between the regular variation, given by a base signal, and the spike process which accounts for the unusual big variation of the spot price. We propose two factor model of spot price dynamics, and thus we have the following form of our spot price model:

$$S_t = e^{\mu(t)} \left[Y_t^{(1)} + Y_t^{(2)} \right] \tag{1}$$

where $\mu(t) : \mathbb{R}_+ \to \mathbb{R}$ is a deterministic function, which represents the seasonality of the electricity spot price.

We assume that the electricity spot price for normal variations follows the Ornstein-Uhlenbeck (OU) process which is governed by the following stochastic differential equation:

$$\mathrm{d}Y_t^{(1)} = -\lambda_1 Y_t^{(1)} + \sigma \mathrm{d}B_t \tag{2}$$

where λ_1 is the mean-reverting rate of return of the spot price, $(B_t)_{t\geq 0}$ is the standard Brownian motion, and σ is the constant volatility. We now introduce an idea of modeling the jumps in the electricity spot process by using Lévy-type Cox process, which we briefly explain:

Define, $\{t_m, m \geq 1\}$ be the sequence of successive times at which "jumps" or "spikes" occur. We refer spikes as jump events which increases electricity spot price, and the subsequent mean reverting behavior. For each time a jump occurs the spot price increases by Z, a random variable with probability density function $f(\cdot)$. As time passes, the spot price decreases until next jump occurs. Thus $Y_t^{(2)}$ satisfies the following stochastic differential equation:

$$\mathrm{d}Y_t^{(2)} = -\lambda_2 Y_t^{(2)} + \mathrm{d}L_t \tag{3}$$

Here L_t is the stochastic point process with underlying Poisson process $N(t)$ with intensity function $I(t)$. The intensity function is time-dependent in order to capture the seasonality in occurrence of the spikes. If $\{t_m\}$ are the times at which the jumps occurs in the interval $(0, t)$, then the corresponding spike process is given by

$$Y_t^{(2)} = y_0 e^{-\lambda_2 t} + \sum_{j=1}^{N_t} Z_j e^{-\lambda_2 (t - t_j)} \tag{4}$$

where Z_j are independent and identically distributed random variables with probability density function $f(\cdot)$. The intensity function $I(t)$ and the probability density function of random variables Z_j are discussed in the following section.

3 Calibration of the Model

The calibration of the model described in the previous section, $S_t = e^{\mu(t)} Y_t$, for the case of the Colombian market data, is done by adopting the methodology proposed in [3, 11], which will be described in this section. We considered the daily 19:00 PM price of energy [COP / kWh] in the "Bolsa de Energía". Recall that the price in "Bolsa de Energía" is fixed per hour the previous day.

The strategy for the calibration consists in identifying the three components, the deterministic seasonal trend and the two stochastic factor processes, in the particular series. We will follow the next steps:

1. Filter the deterministic seasonal trend, $\mu(t)$, from the series.
2. From the stochastic part, $X_t = Y_t^{(1)} + Y_t^{(2)}$, filter each component from the other, in the series.
3. Estimate the parameters of the diffusion process, $Y_t^{(1)}$.
4. Estimate the parameters of the Lévy-type Cox process, $Y_t^{(2)}$.

We recall from the previous section that the functional form is:

$$\mu(t) = \alpha + \beta t + \gamma \cos(\varepsilon + 2\pi t) + \delta \cos(\zeta + 4\pi t)$$

which captures yearly and semiannual seasonalities, where the unit of measure is a year. The parameters are estimated using the nonlinear least square estimator, implemented in the Optimization Toolbox of MATLAB. The estimated values are: $\alpha = 3.7530$, $\beta = 0.0889$, $\gamma = 0.0835$, $\varepsilon = -0.0595$, $\delta = -0.1081$, and $\zeta = 0.6219$, and in Figure 2 we present the series of the price and the log-price with their respective trends and the deseasonalized series.

We proceed to estimate the parameters of the process $X_t = Y_t^{(1)} + Y_t^{(2)}$ using the deseasonalized series, $e^{-\mu(t_i)} S_{t_i}$. The parameters in this case are λ_i the mean reversion rates of each factor, and the parameters of each of the backward driving process (BDP), L_t and $W_t = \sigma B_t$, of the processes $Y_t^{(1)}$ and $Y_t^{(2)}$, respectively.

Using the autocorrelation structure of the process X_t we estimate the parameters λ_i minimizing the L^2-distance between the empirical and the theoretical autocorrelation function. For the parameters λ_i is used the fact that the sum of OU processes autocorrelation function, $X_t = Y_t^{(1)} + Y_t^{(2)}$, is given by $\rho(h) = \tilde{w}_1 e^{-h\lambda_1} + \tilde{w}_2 e^{-h\lambda_2}$, where $\tilde{w}_i = \frac{\mathrm{Var} L_t^{(i)}}{\mathrm{Var} L_t^{(1)} + \mathrm{Var} L_t^{(2)}}$. Minimizing the L_2-distance between the theoretical and the empirical autocorrelation an estimation of the parameters $\tilde{w}_i$ and λ_i is obtained. In this case, $\tilde{w}_1 = 0.5491$, $\lambda_1 = 0.0056$, $\tilde{w}_2 = 0.4509$, and $\lambda_2 = 0.1454$. The comparison of the theoretical and empirical autocorrelation functions is presented in Figure 3.

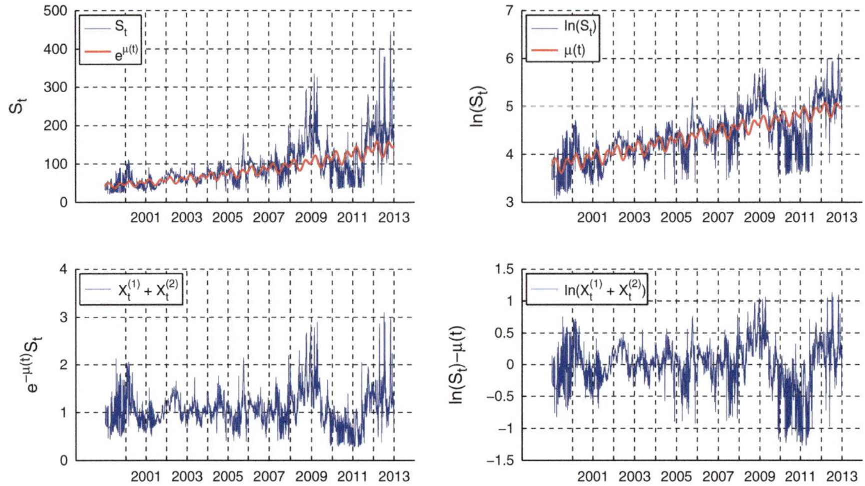

Fig. 2 Price and log-price of electricity

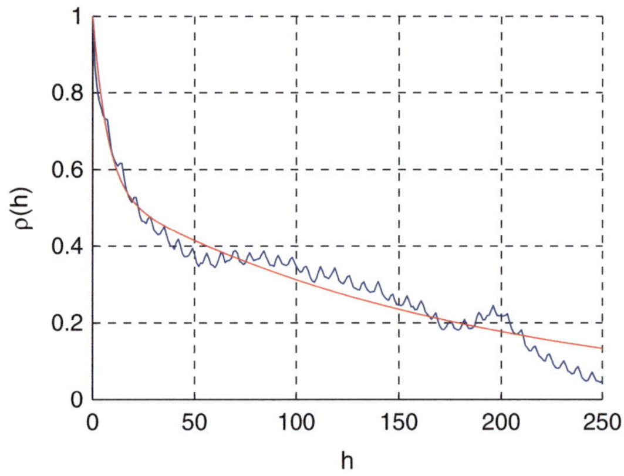

Fig. 3 Empirical and theoretical autocorrelation functions

The next step is to separate the spikes process, $Y_t^{(2)}$, from the base signal $Y_t^{(1)}$. In [3, 11], for the identification of the spike process it is assumed a finite activity driven Lévy process for the OU process that describes the spikes, as in our case. Then, the trajectories have the functional form:

$$f(t) = \sum_{i=1}^{M} \alpha_i 1_{t \geq \tau_i} e^{-(t-\tau_i)\lambda_2}.$$

The methodology to filter the spikes, proposed in [11], is described now. Taking the observations $\{X_{t_j}\}_{j=1}^{n}$ and the transformation $\Delta X_{t_j} = X_{t_j} - e^{-\lambda_1} X_{t_{j-1}}$ we have that:

$$
\begin{aligned}
\Delta X_{t_j} - \Delta Y_{t_j}^{(2)} &= \Delta Y_{t_j}^{(1)} \\
&= Y_{t_j}^{(1)} - e^{-\lambda_1} Y_{t_{j-1}}^{(1)} \\
&= Y_0 e^{-\lambda_1 t_j} + \int_0^{t_j} e^{-\lambda_1 (t_j - s)}\, \mathrm{d}W_s \\
&\quad - e^{-\lambda_1} \left[Y_0 e^{-\lambda_1 t_{j-1}} + \int_0^{t_{j-1}} e^{-\lambda_1 (t_{j-1} - s)}\, \mathrm{d}W_s \right] \\
&= \int_{t_{j-1}}^{t_j} e^{t_j - s}\, \mathrm{d}W_s
\end{aligned}
\tag{5}
$$

which is normal since $W_t = \sigma B_t$ is a Brownian motion. If we assume that indeed we have n discrete observations of the process X_{t_j}, we proceed to estimate the values M, $\{\alpha_i\}_{i=1}^M$, $\{\tau_i\}_{i=1}^M$ for the realization of the trajectory of the jump, given by $f(t)$, our spikes trajectory. Assuming a number of jumps M, since $\left\{\Delta Y_{t_j}\right\}_{j=2}^n$ is a sequence if i.i.d. normal variables, it is then possible to use the Maximum-Likelihood estimator given by:

$$
\arg \min_{(\alpha_j, \tau_j)} \sum_{j=2}^{n} \left(\Delta X_{t_j} - \Delta f(t_j) \right)^2 .
$$

A simple way to approximate the solution of this problem is to place one by one the spikes of the trajectory $f(t)$. The process is applied iteratively until the magnitude of the spikes is less than some threshold. In [11] this method is called the "Hard threshold." The idea goes as follows.

For each $\tau_1 \in \{t_i\}_j^n$, it is possible to compute the optimal jump α_1 given by

$$
\hat{\alpha} = \frac{\Delta X(\tau_1) + \sum_{j=\tau_1+1}^{n} \Delta X(t_j) \tilde{f}(t_j)}{1 + \sum_{j=\tau_1+1}^{n} \left[\tilde{f}(t_j) \right]^2}
$$

where $\tilde{f}(t_j) = e^{-\lambda_2 (t_{j-1} - \tau_1)} \left[e^{-\lambda_2} - e^{-\lambda_1} \right]$, for $t_j > \tau_1$. The chosen value of t_1 which minimizes the mean square error of the expression. Once we have chosen the parameters $(\hat{\alpha}_1, \hat{\tau}_1)$, the same method is applied to the process $X(t) - f_1(t)$, where $f_1(t) = \alpha_1 \mathbb{1}_{t \geq \tau_1} e^{-\lambda_2 (t - \tau_1)}$, and so on, until the threshold is reached. Figure 4 reproduces the original deseasonalized series, the series without spikes, and the series of spike.

For the parameters estimation for the base signal, the assumed Gaussian background driven Lévy process makes the estimation easier. With this assumption

$$
Y_{t_j}^{(1)} - e^{-\lambda_1} Y_{t_{j-1}}^{(1)} = \int_{t_{j-1}}^{t_j} e^{-\lambda_1 (t_j - s)}\, \mathrm{d}Z_s
$$

is normal, and then the discrete process follows a Gaussian $AR(1)$ process and the parameters are estimated using the Maximum-Likelihood estimator, which in this

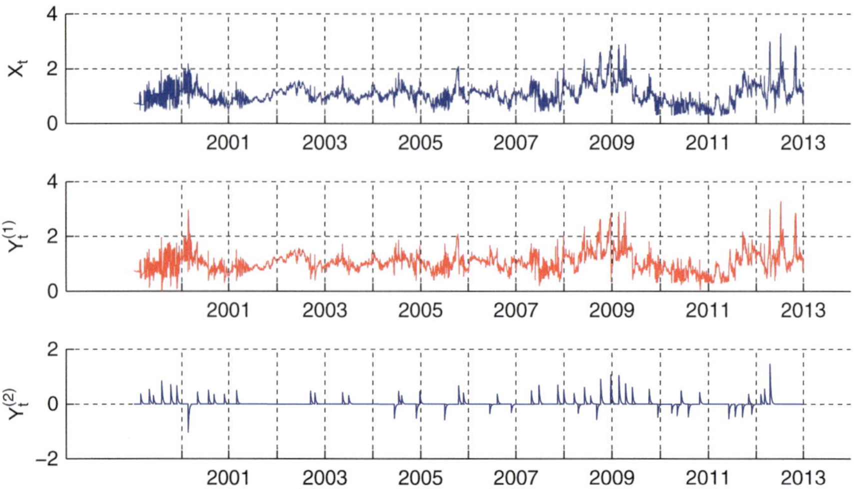

Fig. 4 The first graphic is the deseasonalized series, X_t. The second graphic is the base signal, $Y_t^{(1)}$, and the third graphic is the process of spikes, $Y_t^{(2)}$

case produce the results: $e^{-\lambda_1} = 0.8478$ and $\tilde{\sigma}^2 = 0.03888$. The variance $\tilde{\sigma}^2$ is related with the variance of the B.M., σ, by $\sigma^2 = \dfrac{2\lambda_1\tilde{\sigma}^2}{1-e^{-2\lambda_1 h}}$, where h is the time difference between the observations, in this case $1/365$. So $\sigma^2 = 14.1914$ and so $\sigma = 3.7671$.

According to the different empirical studies considered, there exists evidence of seasonal behavior of spikes. Because of this reason, the Poisson random measure is chosen to describe the jump process associated with the OU process which describes the spikes is taken non-homogeneous. In Figure 5 the spikes distribution within the year is presented. In the left part are the spikes detected and in the right side are considered only the 37 greatest spikes. The Poisson random measure over $[0,t) \times \mathbb{R}$ is defined by the measure $\nu(\,dt \otimes \,dx) = I(t; t_0, \theta, d) \otimes f_X(x)$, where

$$I(t; t_0, \theta, d) = \theta \left[\frac{2}{1 + |\sin(\pi(t - t_0)/k)|} - 1 \right]^d$$

captures the inhomogeneous intensity of spikes in the year, that is, the seasonalities, and $f_X(x)$ is the distribution density of the spikes size. In Figure 6 different configurations of the function $I(t; t_0, \theta, d)$ are represented.

The parameter $1/k$ determines the number of points of accumulation of spikes during one year, in the example $1/k = 3$ implies 3 expected spikes during the year. The parameter d specifies the dispersion around the central point of the spikes. For larger values of d the dispersion is smaller, as Figure 6 shows. In our case, it could be argued that the spikes are accumulated around December and that the dispersion is high, and accordingly were assumed the values $k = 1$ and $d = 0.5$, Figure 7.

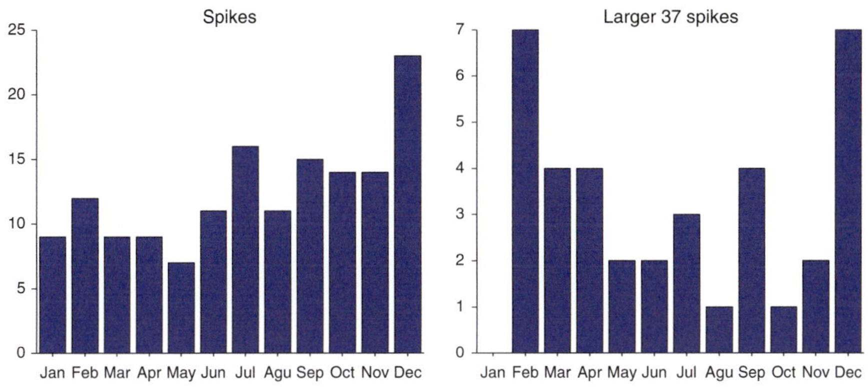

Fig. 5 Spikes distribution

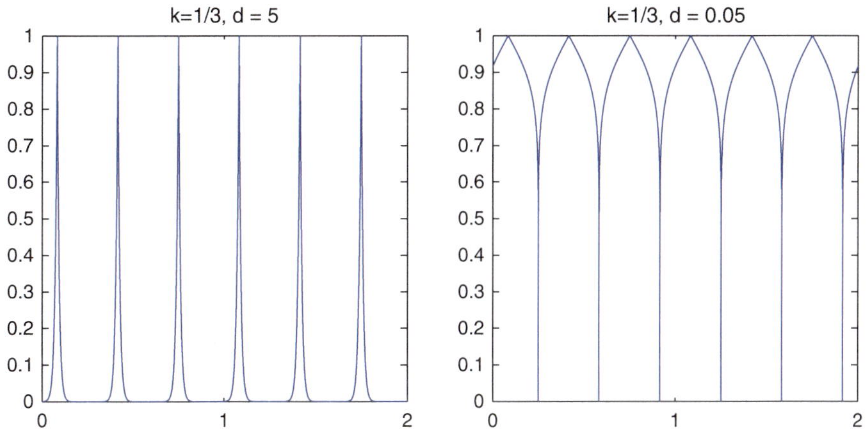

Fig. 6 $I(t; t_0, \theta, d)$

The scale parameter θ, which says how many spikes occur in the year, was estimated according to the Maximum-Likelihood estimator given in [11], given t_0 and d

$$\theta^* = \arg\max_{\theta} \left(\sum_{i=1}^{N_T} \log I(\tau_i; \theta, d) - \int_0^T I(\tau_i; \theta, d)\, \mathrm{d}t \right)$$

which in our case gives

$$\theta^* = \frac{N_t}{\int_0^T h(t)^d \, \mathrm{d}t} = 6.3760$$

where $h(t) = \dfrac{2}{1 + |\sin(\pi(t - t_0)/k)|} - 1.$

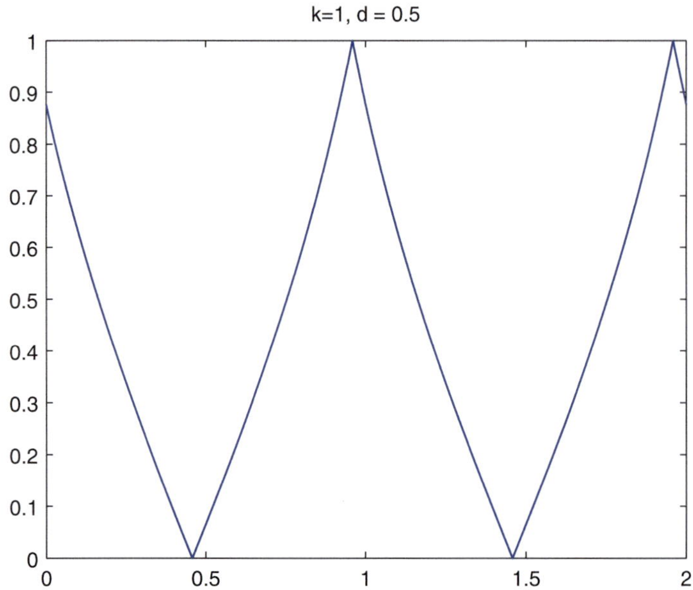

Fig. 7 $I(t, t_0, \theta, d)$

For the estimation of the spikes magnitude distribution in [11], we have used a Pareto distribution $\Pr(Z \geq z) = \left(\frac{z}{z_0}\right)^{\alpha}$, for $z \geq z_0$. In our case seems not to be a good election since the magnitude of the spikes in this case is not very large and the empirical cumulative distributions do not seem to be a straight line, as a Pareto distribution would imply, $\ln\left(\mathbb{P}(Z \geq z) = \alpha \ln(z_0) - \alpha \ln(z)\right.$, see Figure 8. In this case was adjusted a Gamma distribution, $\Gamma(a, b)$, to the spikes magnitude. The sample mean is $\bar{\alpha} = 0.3128$ and the sample standard deviations $\sigma_{\alpha} = 0.1543$, so the parameters in the Gamma distribution are $a = 4.11$ and $b = 0.076$, see Figure 9.

We now complete this section by summarizing the specification of the proposed model for the electricity spot price dynamics S_t and estimated values of the parameters from the calibration.

$$S_t = e^{\mu(t)} \left[Y_t^{(1)} + Y_t^{(2)} \right]$$

$$dY_t^{(1)} = -\lambda_1 Y_t^{(1)} + \sigma dB_t$$

$$dY_t^{(2)} = -\lambda_2 Y_t^{(2)} + dL_t$$

with B_t a standard Brownian motion, and $L_t = \sum_{i=1}^{N_t} Z_i$ a non-homogeneous compound Poisson with intensity

Fig. 8 Empirical Pareto CDF in log–log scale

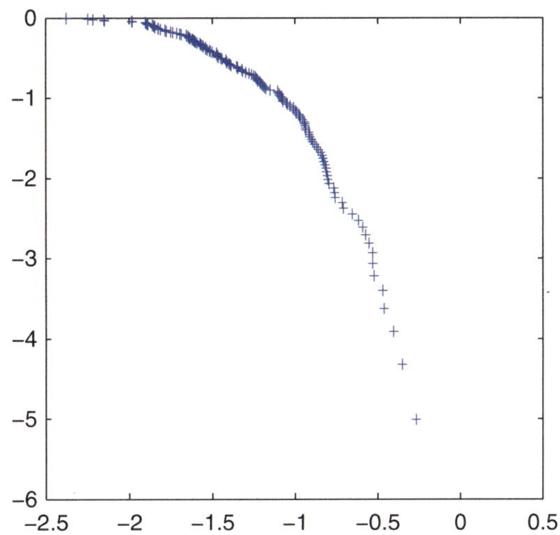

Fig. 9 Empirical and theoretical spikes magnitude distribution

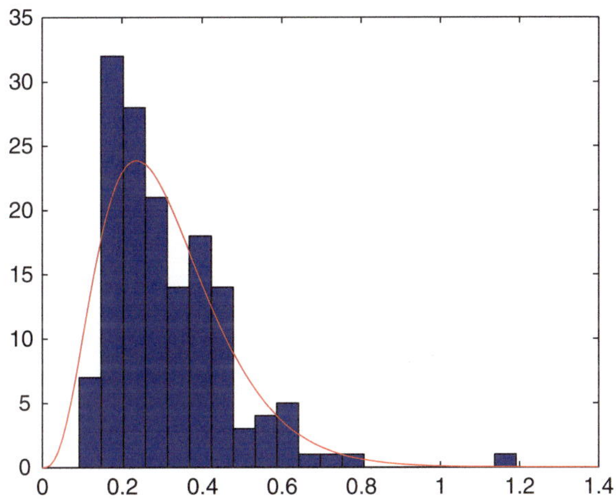

$$I(t; t_0, \theta, d) = \theta \left[\frac{2}{1 + |\sin(\pi(t - t_0)/k)|} - 1 \right]^d$$

and distribution of jumps given by a Gamma distribution, $\Gamma(a, b)$. Finally, the seasonal trend is

$$\mu(t) = \alpha + \beta t + \gamma \cos(\varepsilon + 2\pi t) + \delta \cos(\zeta + 4\pi t).$$

The concrete parameters are:

- For the seasonal trend: $\alpha = 3.753$, $\beta = 0.0889$, $\gamma = 0.0835$, $\varepsilon = -0.0595$, $\delta = -0.1081$, and $\zeta = 0.6219$.

- The mean reversion rates $\lambda_1 = 0.0056 * 365$ and $\lambda_2 = 0.4509 * 365$.
- For the volatility of the Brownian motion for the first component $\sigma = 3.7671$.
- The parameters for the jump intensity are: $\theta = 6.376$, $d = 0.5$, $k = 1$, $t_0 = 11/12$.
- Finally, for the distribution of the jumps $a = 4.11$ and $b = 0.076$.

With the above-said estimated values of the model, we present in Figure 10 a few simulated and original series for the spot price process $X_t = e^{-\mu(t)} S_t$.

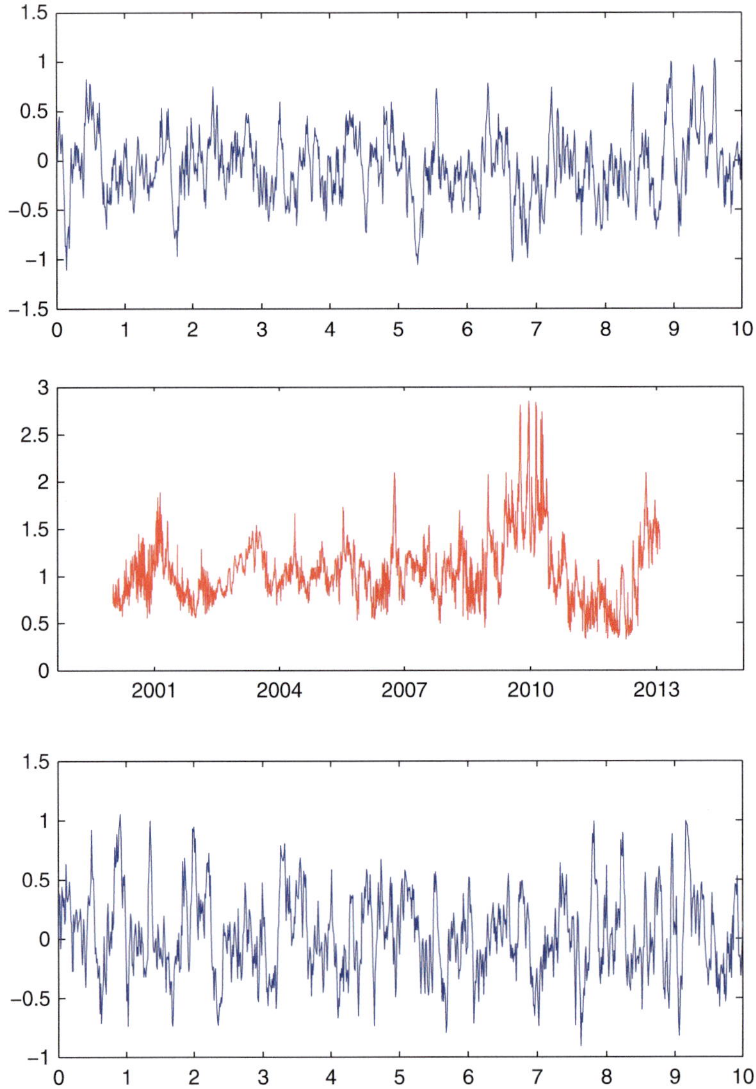

Fig. 10 Simulations and original series

4 Conclusion

In this chapter, we have examined the Lévy-type Cox process with mean reverting tendency of the electricity spot prices for the Colombian Market. A calibration methodology has been implemented and analyzed the empirical performance of the electricity spot price. The simulations suggest a good performance for the proposed model. However, the results computed here are compared with those obtained in the German and Nord market; it is clear that the spikes in Colombian market data are less sharp than in those cases. This has been seen from the comparison between the median of the spikes 0.2810, with its maximum value $1.191, 1.191/0.2810 = 4.2$ which is considerably less than in German and Nord pool where this relation is above 10. We have also seen the better statistical performance of Gamma distribution with respect to the Pareto distribution, which indicates that the magnitudes are smaller in the Colombian market data. In future, we plan to consider a regime-switching behavior in the sport prices for Colombian Market data, in which the spot price alternates from periods of low activity of "base regime" to periods of high activity or "spike regime."

Acknowledgements The authors would like to thank the referees for their comments and suggestions, which have improved the presentation of this chapter.

References

1. Barndorff-Nielsen, O.E., Shephard, N.: Non-Gaussian Ornstein-Uhlenbeck-based models and some of their uses in financial economics. J. R. Stat. Soc. **63**(2), 167–241 (2001)
2. Benth, F.E., Kallsen, J., Meyer-Brandis, T.: A non-gaussian ornstein-uhlenbeck process for electricity spot price modeling and derivatives pricing. Appl. Math. Finance **14**(2), 153–169 (2007)
3. Benth, F.E., Kiesel, R., Nazarova, A.: A critical empirical study of three electricity spot price models. Energy Econ. **34**, 1589–1616 (2012)
4. Botterud, A., Kristiansen, T., Ilic, M.: The relationship between spot and future prices in the nord pool electricity market. Energy Econ. **32**(3), 967–978 (2010)
5. Cartea, A., Figueroa, M.G.: Princing in electricity markets: a mean reverting jump diffusion model with seasonality. Appl. Math. Finance **12**(4), 313–335 (2005)
6. Deng, S.: Stochastic models of energy commodity prices and their applications: Mean-reversion with jumps and spikes. J. Regul. Econ. **19**(3), 239–270 (2001)
7. Furio, D., Meneu, V.: Expectations and forward risk premium in the spanish deregulated power market. Energy Policy **38**, 784–793 (2010)
8. Geman, H., Roncoroni, A.: Understanding the fine structure of electricity prices. J. Bus. **79**(3), 1225–1261 (2006)
9. Kluppelberg, C., Meyer-brandis, T., Schmidt, A.: Electricity spot price modeling with a view towards extrme spike risk. Quant. Finan. **10**, 963–974 (2010)
10. Lucia, J., Schwartz, E.: Electricity prices and power derivatives: evidence from the nordic power exchange. Rev. Deriv. Res. **5**(1), 5–50 (2002)
11. Meyer-Brandis, T., Tankov, P.: Multi-factor jump-diffusion models of electricity prices. Int. J. Thoretical Appl. Finance **11**(5), 503–528 (2008)

12. Redl, C., Haas, R., Huber, C., Bohm, B.: Price formation in electricity forward markets and the relevance of systematic forecast errors. Energy Econ. **31**, 356–364 (2009)
13. Rosiński, J.: Series representations of Lévy processes from the prespective of point processes. In: Barndorff-Nielsen, O.E., Mikosch, T., Resnick, S.I. (eds.) Lévy processes - Theory and Applications. Birkhäuser, Boston (2001)
14. Schwartz, E.: The stochastic behaviour of commodity prices: implications for valuation and hedging. J. Finan. **52**, 923–973 (1997)

Using Value-at-Risk (VaR) to Measure Market Risk of the Equity Inventory of a Market Maker

Argyn Kuketayev and James Beatty

Abstract We propose a simple approach to using value-at-risk (VaR) to measure market risk within the equity inventory of a market making entity, a task which presents several challenges specific to the market making function. Market makers constantly stand ready to buy and sell shares to market participants. In doing so, they inevitably maintain the inventory of shares, a portfolio of sorts, subject to market risk. VaR is a standard tool for measuring the market risk in investment portfolios, so a variety of calculation techniques have been developed over the years. However, the application of VaR to market making inventories requires a few adjustments, for unlike the typical investment portfolio these inventories change rapidly, as if they were rebalanced intra-day. Moreover, the number of unique tickers in the inventory for a given day may routinely list thousands of securities. As a result, at any moment in an inventory there could be hundreds of items with missing historical price data, which makes challenging the application of even the simplest VaR methods. The approach proposed in this paper deals with the rapidly rebalancing portfolio and missing data issues inherent in market making equity portfolios by rescaling portfolio weight to allow for the application of well-known VaR techniques to very large inventories.

Keywords Value-at-risk • Market making • Equity • Portfolio rebalancing

1 Introduction

Market makers serve an important intermediary function in the financial system, supplying liquidity by standing ready to buy or sell securities as orders arrive from the public. Market makers provide trade immediacy to the public by taking any side of a transaction in a given security, and they also serve a role in price discovery by

A. Kuketayev
E*TRADE Financial Corporation, 1806, St Roman Dr, Vienna 22182, VA, USA
e-mail: argyn.kuketayev@etrade.com

J. Beatty (✉)
KPMG 345 Park Avenue, New York, 10154 NY
e-mail: jbbeatty@gmail.com

J.A. Londoño et al. (eds.), *Actuarial Sciences and Quantitative Finance*, Springer Proceedings in Mathematics & Statistics 135, DOI 10.1007/978-3-319-18239-1_2

carefully observing their order flow and adjusting bid-ask quotes accordingly [1,2]. They profit from the bid-ask spread on the items they quote, buying inventory at a lower bid price and selling at a higher ask price. Though their profit derives largely from this bid-ask spread, secondary sources of return to the market maker stem from the ability to time inventory acquisition or drawdown through the order flow and in some cases from commissions on trades [1].

Despite the public role they play, market makers act on their own behalf, buying or selling from their own portfolio using their own capital [3]. They provide immediate liquidity to the public, but in taking an inventory position they become subject to market risk while waiting to connect buyers and sellers [4]. This inventory market risk arises from the possibility that prices may fall before the market maker is able to sell recently acquired inventory, or that prices may rise on short positions, resulting in losses. Market risk also arises from the information asymmetry between market participants—because market makers contribute to price discovery through their view of the order flow and not from an understanding of individual security fundamentals, they stand at risk from informed traders who buy or sell at advantageous prices before market convergence to a new price level, resulting in losses for unbalanced inventory positions [5].

Market makers generally try to adjust inventory through changing their bid-ask spread. The dealer who has amassed too large of a position in a particular security may try to decrease holdings by lowering the ask price at which he will sell that security, and conversely the dealer who wishes to increase holdings in a security will raise the bid price at which he will buy that security. Beyond this, the relationship between inventory and spread may be also determined by other considerations such as the market maker's risk aversion [6]. Set properly, the spread should ensure that inventories remain flat as bid and ask are finely tuned to attract both buyers and sellers in equal amounts. Research has shown, however, that in reality the rate of inventory adjustment can vary substantially from share to share, so at any given moment a market maker's inventory may consist of both large and small positions in a variety of issuances [7]. Thus, a market maker cannot completely eliminate inventory risk.

Faced with the type of risk such as that posed by a market maker's inventory, practitioners often turn to quantitative tools or measures to aid in risk management. Value-at-risk (VaR), commonly known as VaR, is a standard tool for measuring the market risk in investment portfolios. VaR is a statistical measure which estimates the minimum amount of losses in an investment portfolio at given confidence level, over a fixed time horizon [8]. Practitioners may also use expected shortfall, or a conditional value-at-risk (CVaR), as described by Yamai and Yoshiba [9], to measure the amount of expected losses given that losses exceed VaR.

By calculating their portfolio's VaR, market makers have a useful tool to estimate the market risk inherent in their inventory in terms of possible losses. By gaining an understanding of the level and sources of potential losses, they may make efforts to adjust the portfolio inventory to stay within desired risk tolerances through the means available to them, primarily through adjusting the bid-ask spread as discussed above but occasionally through interdealer trades with other market makers as well.

There are many resources available to practitioners willing to implement VaR in practice, such as the standard reference RiskMetrics Classic [10], and a variety of other VaR estimation techniques have been developed throughout the years, as discussed in Jorion [11] and Alexander [12].

2 The Market Making Portfolio and Existing VaR Methods

A market making portfolio distinguishes itself in several ways when compared to a typical investment portfolio. Unlike a typical investment portfolio, the market making inventory is not held to generate returns, because market makers make money on bid-ask spreads rather than investment returns of their holdings. The pure equities inventory would have long and short equity positions hedging each other resulting in the relatively small net position compared to the net long and net short legs, would have nearly net zero beta, and would comprise a nearly delta-neutral portfolio. The chart below, which comes from historical data of a market making entity, illustrates how the long, short, and net positions of a market maker's inventory might change from day to day. Note that the hedging between long and short legs is never perfect—in fact on average this market maker's portfolio is long—so the market maker remains exposed to market risk through an unbalanced portfolio as well as through basis risk (Fig. 1).

Another distinguishing feature of a market making inventory is high daily variability of portfolio composition compared to a typical investment portfolio. As can be seen in the chart blow, drawn from actual data, the number of unique security names held in the portfolio may change greatly day over day, and overall portfolio turnover remains at consistently high levels, near 20 %, with occasional spikes in activity. We define the daily turnover as the ratio of number of shares sold or bought to the number of shares at the beginning of the day (Fig. 2).

Combined, these unique characteristics of the market making portfolio compared to those of a typical investment portfolio present several challenges when implementing VaR. Consider a simple historical simulation method, as discussed in Jorion [11]. In this method it is assumed that the value weights remain constant for each stock i on a given day t, i.e., $w_{it} = w_i$. The value weights for a ticker i with number of shares n_{it} and price p_{it} are defined as follows:

$$w_{it} = \frac{n_{it}p_{it}}{\sum_j n_{jt}p_{jt}}, \tag{1}$$

$$\sum_i w_{it} = 1. \tag{2}$$

The historical portfolio returns R_t happen to be the sums of the weighted stock returns r_{it}:

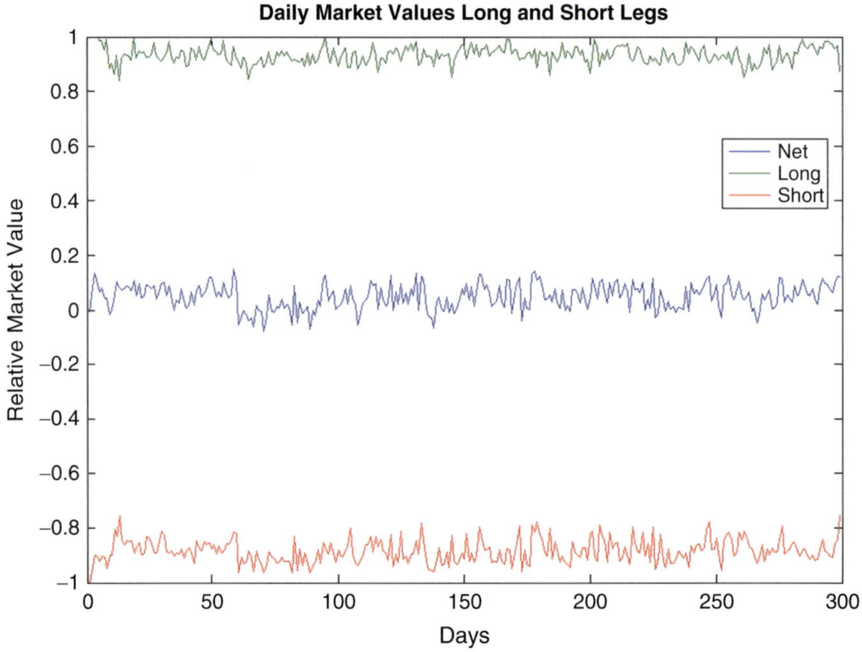

Fig. 1 Daily market values of net position, long and short legs of the market maker's inventory. All values are relative to a long position at the beginning of a period

$$R_t = \sum_i w_i r_{it}, \tag{3}$$

where the stock returns are defined as usual: $r_{it} = \frac{p_{it}}{p_{i,t-1}} - 1 \approx \ln \frac{p_{it}}{p_{i,t-1}}$.

Having the time series of historical portfolio returns R_t, one can estimate the distribution of portfolio returns, then estimate the VaR as its tail at any given confidence α: $\text{VaR}_\alpha = F^{-1}(1-\alpha)$, where $F(.)$ is the inverse probability distribution of returns.

One trouble with implementing this method in a market making portfolio is that because of the large number of held securities (thousands of unique tickers), at any given time t there could be hundreds of stocks for which the prices were not observed for various reasons, e.g. a stock may not be traded very often or it could be a relatively new issue with very short history. In an investment portfolio with a handful of unique tickers various price imputation, interpolation, or removal techniques may be employed (see [13]). However, in an equities inventory with thousands of tickers it is difficult to devise a stable missing data correction methodology, particularly when the data is not *missing at random*, as is often the case with equity price series data. Fig. 3 shows the daily percentage of stocks for

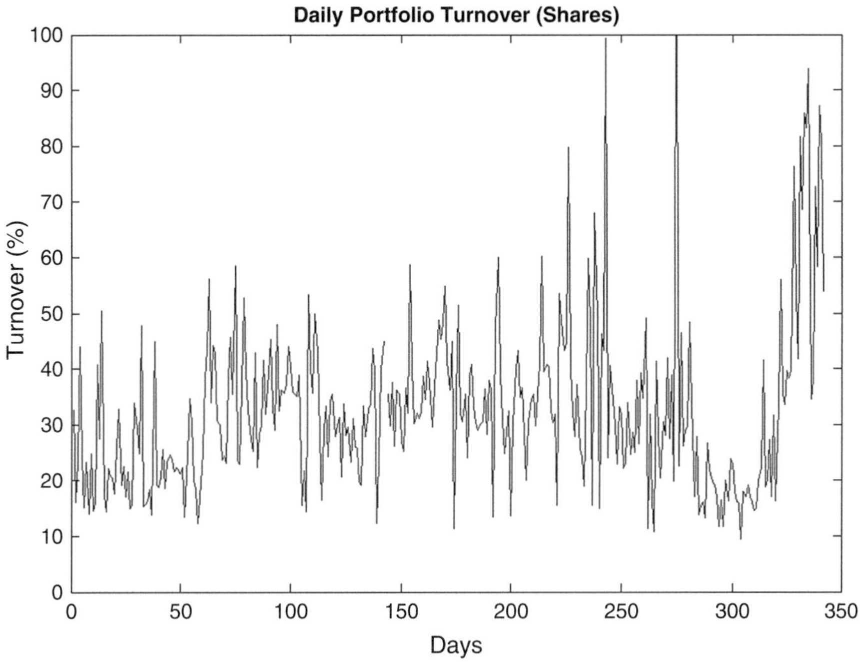

Fig. 2 Daily portfolio turnover rate: the ratio of holdings changed over the total number of shares

which the prices are missing in the inventory of a market maker. On average about 1 % of the stocks—literally, hundreds of unique tickers—do not have price data in this inventory.

Furthermore, consider a risk factor-based VaR methodology, based upon the diagonal model proposed by Sharpe [14]. In this approach, it is assumed that stock returns are correlated with a limited number of time-varying common risk factors f_{jt}:

$$r_{it} = \beta_{ij} f_{jt} + \varepsilon_{it}, \tag{4}$$

where ε_{it} is the diversifiable idiosyncratic risk. For typical investment portfolios this leads to portfolio VaR being equal to a systematic VaR, which is a function of weights and risk factor return distributions, while idiosyncratic risks are usually negligible, see Alexander [12].

For market making inventories, we have the same issue of computing weights for stocks with missing data. However, there is an additional problem: market making inventories are usually delta hedged, i.e., the portfolio betas end up being very small. This in turn makes the idiosyncratic risk magnitudes comparable to systematic risks, so it becomes important to measure these as well.

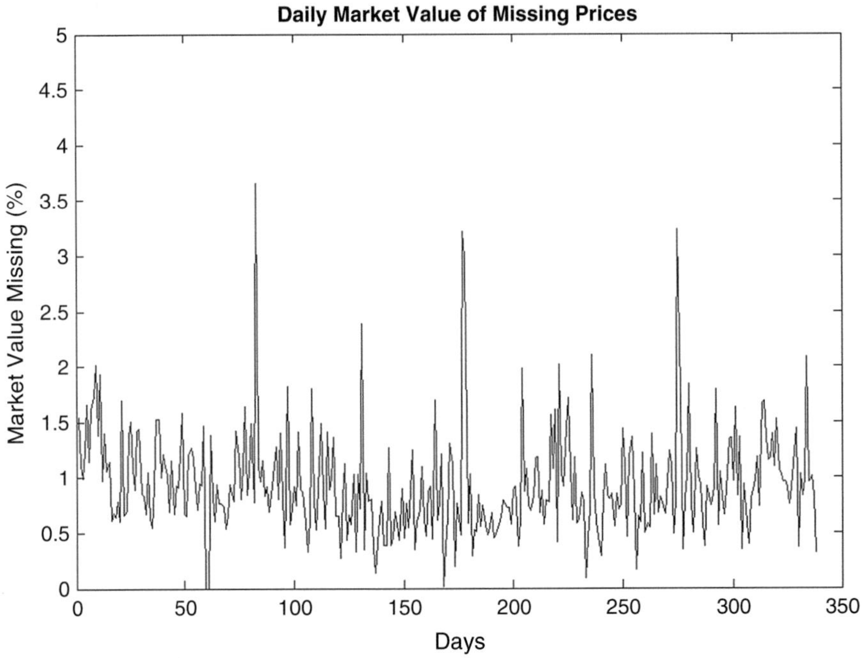

Fig. 3 Daily percentage stock for which the price data is missing

3 New Weight Scaling

To allow calculation of a market making portfolio VaR in methods where daily value weights are assumed to be constant, we propose to rescale the weights as follows. First, set the weights of stocks i with missing data at time t to zero: $\tilde{w}_{it} = 0$, while leaving other weights unchanged: $\tilde{w}_{it} = w_{it}$. Next, calculate the *scaled weights*[1] and apply them to compute the portfolio returns:

$$\omega_{it} = \frac{\tilde{w}_{it}}{\sum_i \tilde{w}_{it}}, \tag{5}$$

$$\sum_i \omega_{it} = 1, \tag{6}$$

[1]Note, that the rescaled weights are time-dependent unlike the original weights, which are constant and capture the current snapshot of the portfolio.

$$R_t = \sum_i \omega_i r_{it}. \tag{7}$$

This rescaling is not just a mathematical trick. It is based on the intuition of the rebalancing of a self-financing portfolio, as discussed by Björk [15]. Consider the portfolio value change from day 1 to day 2:

$$V_2 - V_1 = \sum_i n_{i2} p_{i2} - \sum_i n_{i1} p_{i1}. \tag{8}$$

With a simple algebra it can be represented as follows:

$$\Delta V = \sum_i n_{i1} \left(p_{i2} - p_{i1}\right) - \sum_i p_{i2} \left(n_{i2} - n_{i1}\right). \tag{9}$$

A self-financing portfolio is such that the change in its value comes strictly from the change in the share prices and not from the cash movement in and out of the portfolio due to share sales and purchases, i.e., the following conditions hold:

$$\Delta V = \sum_i n_{i1} \left(p_{i2} - p_{i1}\right), \tag{10}$$

$$\sum_i p_{i2} \left(n_{i2} - n_{i1}\right) = 0. \tag{11}$$

It is easy to see that the proposed weight rescaling is similar to rebalancing a self-financing portfolio: we temporarily redistribute the market value of stocks for which price data is missing for any given day to other stocks in the inventory without adding or removing cash from the value of the portfolio. This procedure is very simple to implement computationally unlike sophisticated imputation techniques used in academic literature.

There is one more complication with market making inventories: they have a relatively large net long and short positions compared to the net of an entire portfolio. In other words, the value of holdings of any given stock i can easily exceed the value of entire portfolio, while for typical investment portfolios this would be unusual. This makes the rescaling unstable when applied to entire portfolio, because in an ideally hedged inventory the sum of weights in the denominator of the Eq. (3) for the scaled weights becomes zero. Therefore, we compute the weights and rescale them for long and short sub-inventories separately. We then obtain the overall returns by combining the returns of sub-inventories.

4 Results and Discussion

Using the scaled weights it is possible to apply a number of standard VaR techniques such as the historical simulation and factor-based approaches mentioned earlier. Note that sometimes we omit the subscript t from the rescaled weights to unclutter the equations for readability purpose.

4.1 Historical Simulation

When implementing the historical simulation approach, we split the market making portfolio into long and short legs:

$$P_0 = \sum_{i \in \text{long}} n_i p_{i0} + \sum_{i \in \text{short}} n_i p_{i0} = P_0^L + P_0^S. \tag{12}$$

By applying a set of daily historical returns r_{it} of stocks i from days t to the current portfolio we simulate the set of portfolio returns of long and short holdings:

$$r_t^L = \sum_{i \in \text{long}} w_i r_{it}, \tag{13}$$

$$r_t^L = \sum_{i \in \text{short}} w_i r_{it}. \tag{14}$$

The weights in Eqs. (13) and (14) are scaled weights as described in Eq. (7). Next, we obtain the set of simulated values of long and short holdings of a portfolio tomorrow using the historical returns:

$$P_{1,t} = P_{1,t}^L + P_{1,t}^S = P_0^L\left(1 + r_t^L\right) + P_0^S\left(1 + r_t^S\right). \tag{15}$$

Here, subscripts 0 and 1 refer to next day's and today's portfolio value; the time index t refers to the historical observation time.

Finally, we get the set of total market value changes—profits or losses—that correspond to days t in the historical period.

$$\Delta P_t = P_{1,t} - P_0 = r_t^L + r_t^S. \tag{16}$$

Because VaR is defined in terms of losses, it is convenient to flip the sign and consider the set of losses $\text{Loss}_t \equiv \{-\Delta P_t\}$ so that a negative market value change is a positive amount of loss. In this method the VaR is simply the $(1 - \alpha)$ percentile of the set of losses $\{\text{Loss}_t\}$

4.2 Risk Factor VaR

In this approach we model the log returns of stocks as:

$$r_{it} = \beta_{oi} + \beta_i r_{xt} + \varepsilon_{it}, \tag{17}$$

where r_{it} and r_{xt} are the returns of stock i and the S&P 500 index (or other suitable risk factor) at time t, ε_{it} is the specific (idiosyncratic) return, β_i is the beta of stock i, and β_{0i} is an intercept, usually called the alpha of stock i.

We note that the next-day return is composed of systemic (risk factor-based) and specific components, and that the total variance of the next-day portfolio returns is a sum of systemic and specific variances, assuming that these portions are uncorrelated with each other ([11], Section 8.2.2):

$$\mathrm{Var}\left[P_{1,t}\right] = \sigma_P^2 + \mathrm{Cov}\left[\varepsilon_{it}\right], \tag{18}$$

where the next-day portfolio value defined as $P_{1,t} = P_0\left(1 + \sum_i w_{it} r_{it}\right)$.

The first term is the systemic portion of the volatility of total portfolio value P_1 defined by the following equations:

$$\sigma_P = \beta_\$ \sigma_x = \left(\beta_\$^L + \beta_\$^S\right)\sigma_x, \tag{19}$$

$$\beta_\$^L = \sum_{i \in \text{long}} n_i p_{i0} \beta_i, \tag{20}$$

$$\beta_\$^S = \sum_{i \in \text{short}} n_i p_{i0} \beta_i, \tag{21}$$

where σ_x is the daily volatility of the S&P 500 index and $\beta_\$$ represents the beta dollars of a portfolio [11].

Under the assumption that risk factor returns—and subsequently stock returns—are from normal distributions, the systemic VaR of the portfolio is computed using a variation of a standard formula:

$$\mathrm{VaR}_\beta = \Phi^{-1}\left(1 - \alpha\right)\sigma_x\left|\beta_\$\right|, \tag{22}$$

where Φ^{-1} is the inverse cumulative standard normal distribution and α is the desired significance level. Note that normal distribution assumption is not essential in this case, and one could employ the empirical probability distributions or suitable distributions from a fat-tailed stable distribution family.

The second term of Eq. (18) is a specific, or idiosyncratic, variance add-on. Here we estimate the error terms with regression residuals:

$$e_{it} = r_{it} - (\beta_{0i} - \beta_i) \, . \tag{23}$$

We then split the position to long and short holdings to compute the specific variance:

$$\text{VaR}_{\text{Spec}} [P_1] = \Phi^{-1} (1 - \alpha) \sqrt{P_0' \Omega P_0}, \tag{24}$$

$$P_0 = \left[P_0^L, P_0^S \right]' , \tag{25}$$

$$e_t = \left[e_t^L, e_t^S \right]' , \tag{26}$$

$$\Omega = \text{Var} \left[e_t \right] . \tag{27}$$

Here, the terms $e_t^L = \sum_{i \in \text{long}} w_{it} e_{it}$ and $e_t^S = \sum_{i \in \text{short}} w_{it} e_{it}$ are the weighted series of residual returns for long and short holdings on days t, P_0 is a matrix of today's market value of the portfolio, where market values of long and short holdings are the components of the column vectors, and Ω is the variance–covariance matrix of the residual returns vector e. Crucially, the weights are not the current weights w_{i0} but are the scaled weights which take into account missing data as described in Eq. (7).

Total VaR is then given by equation:

$$\text{VaR}_{\text{Total}} = \sqrt{\text{VaR}_\beta^2 + \text{VaR}_{\text{Spec}}^2}, \tag{28}$$

here we assume that the idiosyncratic risk is uncorrelated with risk factors. Note that we did need assumptions about correlations between individual stocks.

4.3 Results

In Fig. 4 we demonstrate the daily VaR series calculated using these two approaches. Both methods render very similar results.

We show the systemic and specific components of daily VaR calculated in risk factor-based approach in Fig. 5. As discussed earlier, in contrast with a typical investment portfolio, a market maker inventory's specific risk is much larger than the systematic risk.

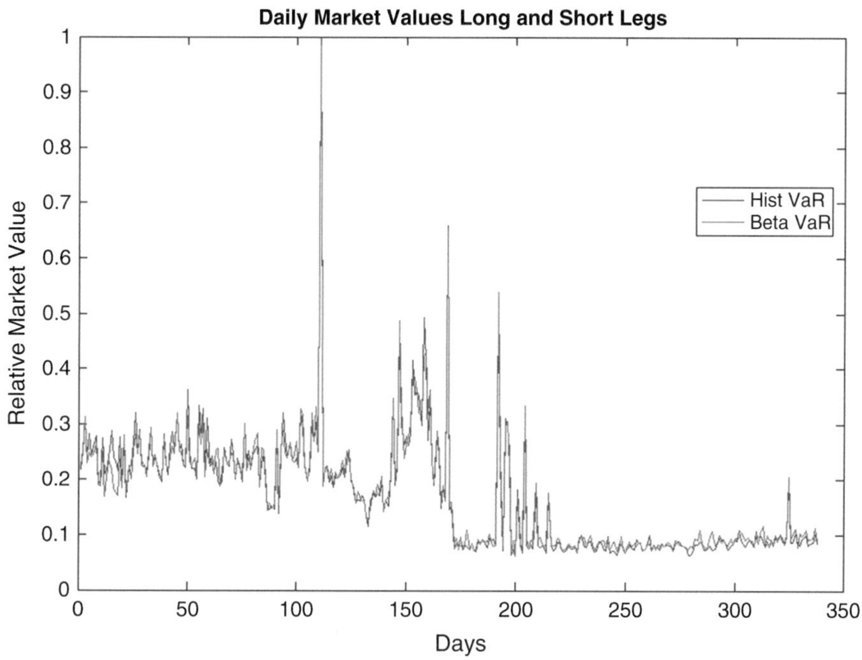

Fig. 4 Daily VaR: historical simulation and risk factor-based approaches. All values are relative to max value of historical VaR in the sample

We conclude that the proposed weight rescaling allows applying well-known VaR methodologies to market making equity inventories. It is based on an idea of rebalancing of self-financing portfolio. The method is computationally efficient and easy to implement. Thus it can be recommended for application in measuring the inventory risk of a market maker.

References

1. Schwartz, R.A., Francioni, R.: Equity Markets in Action. Wiley, Hoboken (2004)
2. Harris, L.: Trading and Exchanges: Market Microstructure for Practitioners. Oxford University Press, New York (2002)
3. Schwartz, R.A., Peng, L.: Market makers. In: Lee, C., Lee, A.C. (eds.) Encyclopedia of Finance. Springer, New York (2006)
4. Grossman, S.J., Miller, M.H.: Liquidity and market structure. J. Financ. **43**(3), 617–633 (1988). doi:10.1111/j.1540-6261.1988.tb04594.x
5. Glosten, L.R., Milgrom, P.R.: Bid, ask and transaction prices in a specialist market with heterogenously informed traders. J. Financ. Econ. **14**(1), 71–100 (1985). doi:10.1016/0304-405X(85)90044-3
6. O'Hara, M., Oldfield, G.S.: The microeconomics of market making. J. Financ. Quant. Anal. **21**(04), 361–376 (1986). doi:10.2307/2330686

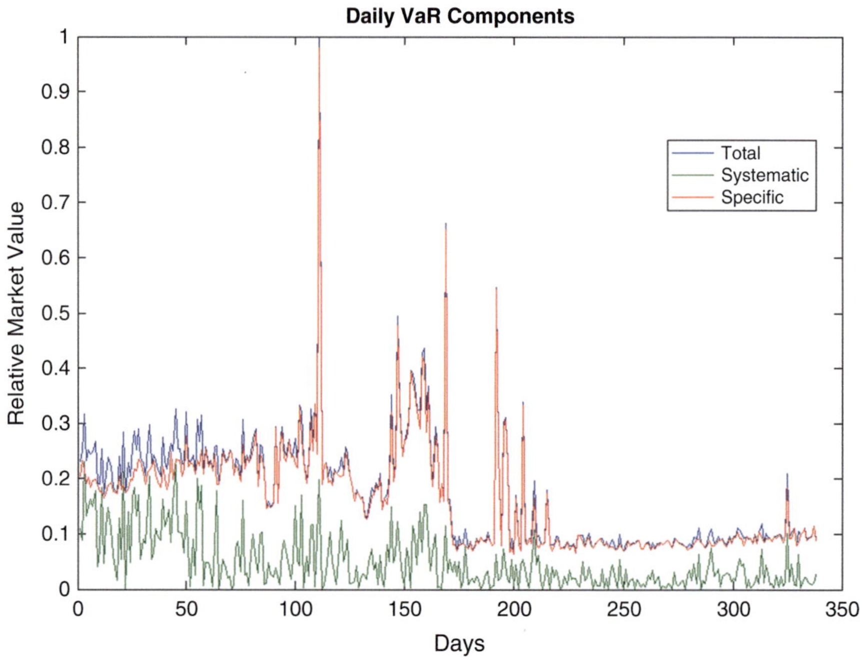

Fig. 5 Components of daily risk-based VaR: total, systematic, and specific

7. Hasbrouck, J., Sofianos, G.: The trades of market makers: an empirical analysis of NYSE specialists. J. Financ. **48**(5), 1565–1593 (1993). doi:10.1111/j.1540-6261.1993.tb05121.x
8. Linsmeier, T.J., Pearson, N.D.: Value at risk. Financ. Analysts. J. **56**(2), 47–67 (2000). doi:10.2469/faj.v56.n2.2343
9. Yamai, Y., Yoshiba, T.: On the validity of value-at-risk: comparative analysis with expected shortfall. Monet. Econ. Stud. **20**(1), 57–85 (2002)
10. Riskmetrics (1996) Technical Document. Morgan Guaranty Trust Company of New York
11. Jorion, P.: Value at Risk: The New Benchmark for Managing Financial Risk. McGraw-Hill Companies, Inc., New York (2006)
12. Alexander, C.: Market Risk Analysis, Value at Risk Models. Wiley, Chichester (2008)
13. Little, R.J.A. Rubin, D.B.: Statistical Analysis with Missing Data. Wiley, New York (1987)
14. Sharpe, W.F.: A simplified model for portfolio analysis. Manag. Sci. **9**(2), 277–293 (1963). doi:10.1287/mnsc.9.2.277
15. Björk, T.: Arbitrage Theory in Continuous Time. Oxford University Press, Oxford (2009)

Reverse Mortgage Schemes Financing Urban Dynamics Using the Multiple Decrement Approach

David Bogataj, Diego Ros McDonnell, and Marija Bogataj

Abstract The objective of this chapter is to introduce flexible reverse mortgage contracts and to consider them in a multiple decrement model. Here, the multiple decrement approach is applied for the first time in housing and urban planning as a new tool for better forecasting of the housing market and housing needs of senior citizens with decreasing functional capacities and therefore changing housing needs on their lifetime horizon. The presented approach enables development of appropriate, more flexible equity release schemes (ERS) than the products known up to now. In the flexible reverse mortgage model developed here we assume that the homeowner can always sell her/his residential property and buy the new residential property for the amount which is of the same value as that of the first house, or lower. It enables better management of cash flows related to the portfolio of residential properties in the context of ERS—the reverse mortgage approach. The schemes are designed to allow elderly homeowners to convert the equity in their homes to income and to keep the possibility to sell their property and move to a more appropriate dwelling under the same contract, which was not the case up to now. Namely, in these newly presented ERS products, homeowners can also keep the mathematical reserves for longevity insurance, based on the initial contract, when they move and sell the initial residential property and buy a new one. They also keep their deferred lifetime annuities if they finally sell the residential property and move to long-term care in a retirement home. In this new approach, the reverse mortgage schemes/longevity insurance are not tied to the single property as it is the case in the existing reverse mortgage products, but rather to the life of the senior homeowner. In the numerical examples, different cash flows associated with

D. Bogataj (✉)
Evropska Pravna Fakulteta (The European Faculty of Law), Delpinova 18/b, Nova Gorica, Slovenia
e-mail: dbogataj@actuary.si

D.R. McDonnell
Universidad Politecnica de Cartagena, Plaza Cronista Isidoro Valverde, Cartagena, Spain
e-mail: diego.ros@upct.es

M. Bogataj
MEDIFAS, Vrtojba—Mejni prehod 6, Sempeter pri Gorici, Slovenia
e-mail: marija.bogataj@guest.arnes.si

J.A. Londoño et al. (eds.), *Actuarial Sciences and Quantitative Finance*, Springer Proceedings in Mathematics & Statistics 135, DOI 10.1007/978-3-319-18239-1_3

different paths of a graph are presented demonstrating the advantages of the flexible reverse mortgage. We show that not only the transaction costs are lower but also that the set of possible choices is larger. According to population aging in Europe, where in 2060 one-third of the population will be more than 64 years old, it is important also for the local authorities and spatial planners to study elderly people's housing needs in order to design a housing system suitable also for them, and therefore to adopt the housing stock to the desired housing consumption. The housing needs of senior citizens may be satisfied if the housing stock is specifically designed to meet their physical, emotional, recreational, medical, and social needs. Recognizing that urban space must be adapted to meet these objectives, the paper will examine the issues regarding the sustainable financing of senior citizens and their changing housing needs in particular.

Keywords Reverse mortgage • Multiple decrements • Equity release scheme • Longevity insurance • Housing market • Urban studies

1 Introduction

The World Health Organization's report "Global Age Friendly Cities" considers population aging and increased urbanization as two major global trends shaping the twenty-first century [11]. The financial crisis in Europe has inflicted heavy losses on older people by increasing the average retirement age, decreasing the value of pensions, decreasing the ECB interest rate, and decreasing returns on assets held by pension funds. The income of senior citizens in Europe has decreased, but the major part of their assets is still equity embedded in their home, i.e., their residential property. Housing is a potential source of both material and environmental well-being [1, 15–18] insofar as it encompasses both housing consumption and investment, especially capital accumulation effects. According to INSEE (www.insee.fr), more than 70 % of senior citizens aged 60–80 live in owner-occupied housing. In the countries with low pension income, like Spain and other Southern and Eastern EU countries, this homeownership rate is over 80 % and has been increasing substantially in the recent 40 years. It means that substantial part of older cohorts is income poor and asset rich. We shall look beyond the crisis to examine the possibility of developing and implementing flexible Housing Equity Release Schemes in European cities [6] as a means of providing improved welfare provision for senior citizens. Equity Release Schemes (ERS) enable older citizens to take cash from the equity built up in their property. These products are targeted at older homeowners who probably have little or no income to make regular repayments or to improve their quality of life. The minimum age is typically 55–60 years; nevertheless, the existing reverse mortgage contracts terminate when the homeowner moves to another dwelling.

Properly developed ERSs could provide better welfare support for senior citizens by:

(a) Enabling easier movement from less appropriate to more appropriate housing units. Existing ERSs in many developed countries do not allow change of dwelling of senior citizens while the ERS contract is still in force. The owners have the only option to sell the property, repay the mortgage, and, with the remaining proceeds, buy a new, smaller, or less valuable property if the price achieved is high enough. Only after the completion of these transactions they are free to close another ERS, but they need to pay a new insurance premium;
(b) Providing low transaction costs when changing their residential property used as collateral, to enable senior homeowners to adopt their housing consumption to their housing requirements due to decreasing functional capacity. Seniors typically adjust their housing consumption to desired or affordable levels by moving, or by investment in home improvements. In case of no transactions costs, a senior would adjust his/her housing immediately whenever desired consumption changed, usually dependent on decreased functional capacity. The costs of searching and moving are substantial [13, 14]. According to theory, in response to a change in housing demand, a senior will decide to search for an alternative dwelling if the expected utility gains from the search outweigh the utility costs of searching, concluding the transaction, and moving [21];
(c) Supporting appropriate supply networks for care: With the disbursements received from the reverse mortgage contract they can cover the costs of additional services needed for a comfortable life in old age;
(d) Ensuring an improved welfare mix at low interest rates: With the disbursements received from the reverse mortgage contract they can cover the costs of different amenities or superior health care, but it all depends on the level of the interest rate, which is an important parameter in ERS products.

The aging of European population is progressing rapidly, while influencing the patterns of urban growth [20] due to different needs of senior citizens regarding housing and contacts with the environment. The reasons are the decreasing functional capacities of senior citizens, who wish to stay in their own property in most cases as long as possible. [17], for example, found a positive correlation between home ownership and both happiness and physical health, although it is important to differentiate housing-specific satisfaction from life satisfaction, or happiness in general. Though the housing industry has substantially focused on senior citizens in the recent 20 years, this was often not considered as an important component of changing urban structures [8, 10]. However, considering the pace of population change, it is necessary to develop new economic and social conditions in the urban structures and new kinds of facility management, especially when we wish to support a longer aging in place. It is important also for the local authorities and spatial planners to study elderly people's housing needs in order to design a housing system suitable for them. The housing needs of senior citizens may be satisfied if the housing stock is specifically designed to meet their physical, emotional, recreational, medical, and social needs. The details about housing needs of older cohorts are given in [3]. However, special financial sources may be required to ensure such needs are adequately met. Recognizing that urban space must be adapted to meet these objectives, the paper will examine issues regarding the sustainable financing of senior citizens and their changing housing needs in particular. Here we shall develop a new ERS products and the method how to forecast a possible portfolio of housing assets under reverse mortgages. The subject of the existing ERSs in Europe and globally is single residential property only. The existing reverse mortgage contracts are in force only as long as the owner, i.e., senior citizen, is also the resident of that housing unit, while she/he cannot move to a more appropriate unit when his/her functional capacity decreases, without terminating the ERS contract. If the senior citizen needs to move to a more appropriate dwelling, then the existing ERS contract is terminated and the property is foreclosed after the owner moves out of the property. The holder loses all services and assurances associated with the ERS contract. For example, in USA HECM insurance which also includes the insurance for longevity is terminated after the sale of the property and lapse of the contract: hence the property owners need to close a new HECM contract for the new property. They also need to pay again the full HECM insurance premium, despite the fact that the homeowner's longevity risk was already covered by the first contract. Such additional costs deter property owners from required changes due to reduced functional capacity. According to [2], development of this kind of services is still embryonic in Spain and although homeownership does improve the well-being related to housing, home equity does not exhibit a significant and independent effect in this and many other European countries [1].

To better accommodate the housing needs of senior homeowners we shall develop a new type of ERS schemes which would allow the sale of the existing home and the purchase of a new housing unit—several times if required. We considered how to provide low transaction costs to enable senior homeowners to adopt their housing consumption to their housing requirements due to decreasing functional capacity. **To keep the mortgage contract and longevity insurance in force also when relocating to a more appropriate housing unit is one of the main reductions in transaction costs for seniors having their home under the reverse mortgage.** The current reverse mortgage products like HECM do not allow keeping the longevity insurance and require payment of closing costs each time when the homeowner wants to sell the existing property and purchase a new one.

While developing the flexible reverse mortgage product we have assumed that seniors typically adjust housing consumption to desired or affordable levels by moving, or by investment in home improvements. In case of low transactions costs, a senior homeowner would adjust his/her housing more often, according to desired housing consumption, usually dependent on decreased functional capacity.

The suggested flexible reverse mortgage product is an innovative financial solution that would allow the relocation and change of residential property under the single mortgage contract as many times as needed, without paying additional premium for longevity insurance and additional costs associated with the closing of the ERS contract. The instrument presented in this paper is a combination of the reverse mortgage loan and the longevity insurance in the form of the deferred life-time annuity which covers the period after drawing the full value of residential property. Therefore, we shall present here the special dynamics of mortgage loan repayment, where in case the senior citizen needs to move to the more appropriate dwelling, the existing ERS contract will not be terminated but, rather, remodeled—even though the property is sold after the owner moves out of the previous property—and buys a new one, more appropriate to his reduced functional capacities. Longevity insurance in these financial instruments does not expire when the homeowner wants to change his/her housing unit, but remains in force without additional costs. The term longevity insurance is well known in the actuarial science and actuarial practice and means an annuity contract designed to pay to the policyholder a benefit if she/he survives to a pre-established future age. In current reverse mortgage products it has not been considered as presented here, where homeowners are changing housing units regarding their decreasing functional capacities and the multiple decrement plans. Financial institutions to offer such products also need new innovative risk management models for managing the portfolio of such ERS contracts. For efficient management of the portfolio of ERS contracts, an appropriate model for forecasting the demand and supply of different type of residential units inside a certain city or region should be developed. It will enable us to model properly the cash flows generated by the portfolio.

2 Multiple Decrement Model in Urban Planning

2.1 Decreasing Functional Capacities of Inhabitants

Europeans are living in a watershed time when the main drivers of economy, therefore also financial services have fallen in crisis, and should change their strategies. The decreasing fertility rate has decreased the population growth, while the increasing life expectancy has increased the population growth of older cohorts, also influencing demand in the real estate industry. Nevertheless, the needs of this part of population are different than those of the industrial society in the last two centuries, which should reflect in designing and renewal of towns and cities. Following the results in the IIASA report from the ERD database [9] it is estimated that Europe will have more than 91 million people aged 75 and above by 2050, which will present more than 14 % of the European population, and nearly 82 million people aged from 65 up to 75, which will present nearly 28 % of the European population having 65 years or more. The most problematic area is the Mediterranean Coast, where this percentage is approaching one-third of the total population. Besides the pensions, the most important asset of the average retiree is residential real estate. According to a study of Das Institut für Finanzdienstleistungen E.V., Hamburg (Reifner, 2009) the share of residents that own residential housing units in Europe was between 43 % in Germany and 98 % in Romania in 2007.

In an age-friendly city, policies, facilities, services, settings, and structures support and enable people to age comfortably, enabling the longer independent living [11]. Barrier-free buildings, streets, and other communication facilities enhance the mobility and independence of people with reduced mobility. To stay independent, older people would need to move in such barrier-free buildings when their functional capacities fall under a certain disability threshold [3]. Their home and the outside environment have a major impact on the mobility, independence, and quality of life and have impact on their ability to live independently longer, also aging in place. For example, the condition of pavements has an important impact on the ability to walk safely in the local area. Existence of good customer and facility services also improves independent living of seniors. Having accessible, frequent, and reliable public transport is one of the key factors influencing independent and active aging. Older people who have difficulties using public transport need to have specially adapted means of transport, but such options are not available in all European cities and towns. For the people in cohorts 65+, family size is generally decreasing (their children have their own homes and many lose their partner) and their functional capacities are decreasing as well, as presented in Figure 1. Investments in accessibility improvements can push down the threshold line, while the rehabilitation and the lifestyle of an individual can move the curve of functional capacity upward.

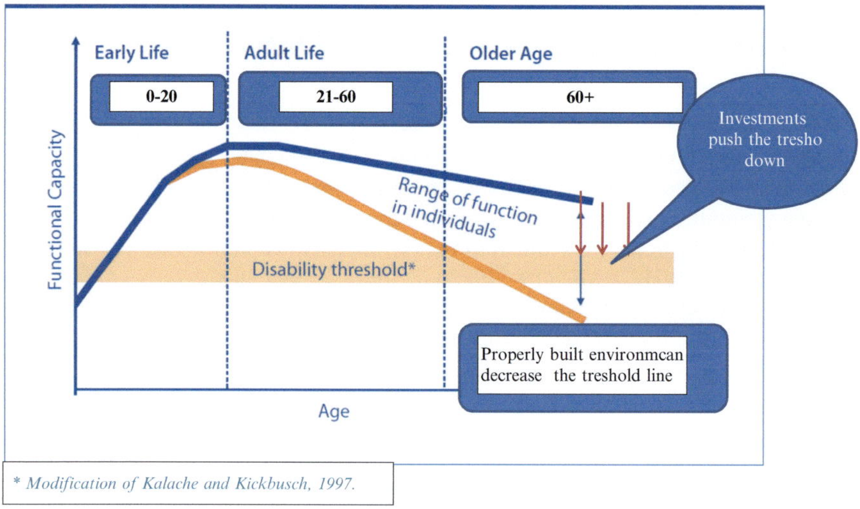

Fig. 1 Decreasing functional capacities of inhabitants

2.2 Move to More Suitable Housing

As demand for housing among the younger population declines and rises among the elderly, many housing units and urban communities will need to be adapted for older populations, which will have spatial implications for provision of welfare. Such trends may produce demographic and spatial pressures for cities, as well as issues regarding social service delivery and organization of retirement communities in the existing town or its surroundings. We must have a plan how to amalgamate communities with the integrative part of the elderly together by investing in facilities which can push down the disability threshold. Indeed, nearly one-third of the housing stock and the built environment needs to be transformed to homes, facilities, and accompanying services for senior citizens. Therefore, substantial financial resources are needed for such renovation activities of towns which cannot be covered by pensions and social benefits for senior citizens only, while local governments' budgets are in poor shape. The asset tied in real property, if properly released, could provide additional sources to better cover those needs. New sources of financing will be required to adapt housing units for aging cohorts, therefore the financial support through equity release mortgages may be a solution. But first of all we have to be able to forecast the dynamic of the needs and effective demand for different categories of buildings and housing units to improve portfolio management. It depends on citizens' migration flow and their curves of functional capacities, and also on the curve of functional capacity of those staying in the study area. While migration flows can be forecast using the gravity model for structured flows regarding age cohorts [5], forecasting of housing supply and demand of different cohorts, regarding decreasing functional capacity, could be improved using the multiple decrement model. Therefore, the multiple decrement models used for critical illness insurance [4, 19] are here applied to housing consumption and household mobility, related to decreasing functional capacities of homeowners and hence connected to the reverse mortgage plans.

2.3 The Model

The multiple decrement models allow for the calculation of the expected tenure in each type of residential property, according to the locational preferences and functional capacities of a homeowner.

For successful forecasting of different housing needs regarding decreasing functional capacities of residents and effective demand, we need to know the probability distribution of $T(x)$ to model the time of transition from the state "alive with full functional capacity" in the family home to a more appropriate housing unit, while facing the loss of family members, functional capacity, or the loss of life. We shall denote by H the set of all different types of housing units listed in Figure 2. In the multiple decrement model with m new types of housing options including also death, there are $m + 1$ states for the transition from one housing to another (smaller and more accessible units for dependent or independent living). We denote the living in family home as state $i = 0$ and the living in any other type of home as state i ?0 and the decrement which requires housing unit of type j as state j, $j = 1, 2, \ldots, m$. The model should describe the probabilities of transition from state 0 to state $j \in H$ at various time points. In the multiple decrement setup, the transitions between any two states i and j, $i > j = 1, 2, \ldots, m$ are not possible, while in multi-state transition we can assume also such transitions [4, 7, 12].

The residents may move among various states such as independent living in cities or segregated retirement communities, temporarily or permanently in a retirement home, or death. The details of the move are presented in Figure 2.

Let us consider a homeowner aged x years. We shall denote this person by (x). We denote the tenure—the future lifetime of the homeowner that he/she will spend in the existing home $i \in H$—by $T_i(x)$. Thus $x + T_i(x)$ will be the age when the homeowner will move out of the existing residence to residence of type $j \in H$, which means a more appropriate dwelling or death. The future tenure in the residential property of type i, $T_i(x)$ is a random variable with a probability distribution function

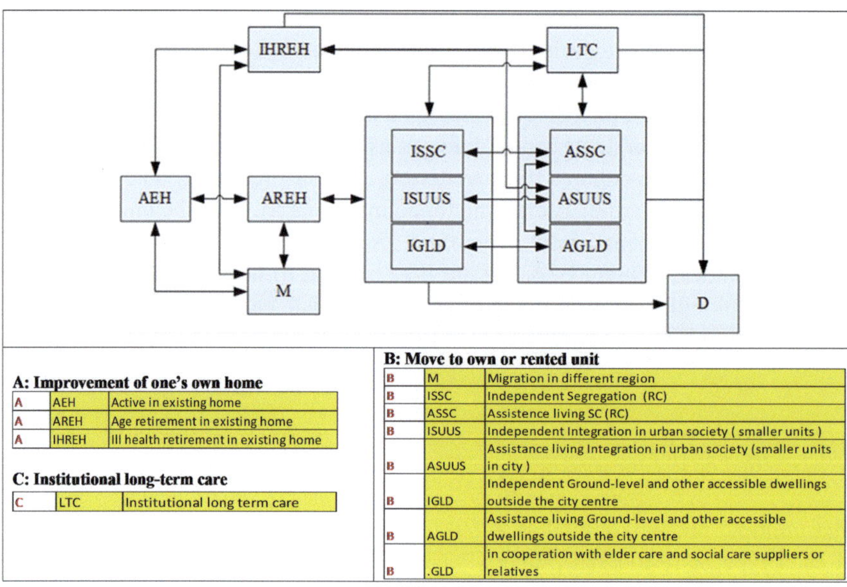

A: Improvement of one's own home

A	AEH	Active in existing home
A	AREH	Age retirement in existing home
A	IHREH	Ill health retirement in existing home

C: Institutional long-term care

C	LTC	Institutional long term care

B: Move to own or rented unit

B	M	Migration in different region
B	ISSC	Independent Segregation (RC)
B	ASSC	Assistence living SC (RC)
B	ISUUS	Independent Integration in urban society (smaller units)
B	ASUUS	Assistance living Integration in urban society (smaller units in city)
B	IGLD	Independent Ground-level and other accessible dwellings outside the city centre
B	AGLD	Assistance living Ground-level and other accessible dwellings outside the city centre
B	.GLD	in cooperation with elder care and social care suppliers or relatives

Fig. 2 The move or migration of residents among various states
$\mathbf{i \to j; ; i \in H, j \in H}$

$$G_i(t) = Pr(T_i \leq t),\ t \geq 0,\ i \in H \tag{1}$$

Function $G_i(t)$ represents the probability that the homeowner will die (or need to move into a more appropriate home for the elderly) within t years, for any fixed t.

We assume that at $G_i(t)$, the probability distribution of T_i is known. We also assume that $G_i(t)$ is continuous and has probability density $g_i(t) = G_i'(t)$. Thus one may describe

$$\begin{aligned} g_i(t)dt &= Pr(t < T_i < t + dt, i \in H) \\ g_i(t) &= \sum_{j=1}^{m} g_{i,j}(t) \end{aligned} \tag{2}$$

where (2) describes the probability that the homeowner will move from housing unit i in the infinitesimal time interval from t to $t + dt$.

If change in functional capacity of homeowner (decrement) occurs in time t, the conditional probability of reallocation to housing unit of type j in the case that he/she is no longer able to stay at home is represented by conditional probability

$$\Pr(H_{\Gamma_i} = j \,|\, T = t, I = i) = \frac{g_{i,j}(t)}{g_i(t)}; \quad I \subset H \tag{3}$$

Therefore the probability that a homeowner aged x in state i will move into housing unit of type j within t years is denoted by symbol ${}_tq_{j,x}(i)$. Here j is one of the child nodes of i ($j \in \Gamma_i \subset H$). Similar to Gerber's derivation (1997) for one specific initial state, we can write:

$$\begin{aligned} {}_tq_{j,x+s}(i) &= \int_s^{s+t} g_j(i,z)dz / \left[1 - G_i(s)\right] \\ {}_tp_x(i) &= 1 - G_i(t) \end{aligned} \tag{4}$$

Here ${}_tp_x(i)$ denotes the probability that a homeowner aged x will stay in his/her residential unit t years and he/she will move at the age $x + t$ to housing unit of type j (see the parallel consideration in [7], for multiple decrement in a set of health statuses). The values of (4) can be observed by city statistics or regional statistics through relative frequencies of these transitions [12] and by studying how purchasing power (also available financial products) influences effective demand on the housing market. In Figure 2, the graph starts at initial state $i = AEH = 0$ and we can observe all possible paths from AEH through some of eleven child nodes $j \in H$, which enable different transitions, from the possibility of migrating out of the region or retiring in an existing home, to moving to a more appropriate housing unit designed for dependent (assisted) or independent living, and further to the unit which provides long-term care in an institutional environment and/or death.

The transitions: ISSC $\leftrightarrow$ ASSC, ISUUS $\leftrightarrow$ ASUUS, and IGLD $\leftrightarrow$ AGLD do not need to be directed and do not necessarily require the change of the residential unit, but could mean just a changed possibility for services of ambient assisted living and could have different levels of services regarding the functional capacity of the owner and the disability threshold of the environment. Services could be sometimes needed and sometimes not.

In our model, we are dealing with the curtate lifetime of a homeowner (x) who is residing in housing unit of type i. If we define by $K_i = [T_i]$ the number of completed future years lived in housing unit of type i by person (x), it means the curtate future tenure of homeowner (x) in housing unit of type i before he/she moves in any successor- housing unit of type j. The probability distribution of the integer-valued random variable K_i is given by

$$\Pr(K_i = k) = \Pr(k \leq T_i < k + 1) = \sum_{j \in \Gamma_i} {}_k p_x(i) q_{j,x+k}(i) \tag{5}$$

where $\sum_{j \in \Gamma_i}$ means the sum over all successors j of the node i, $\Gamma_i \subset H$, in the graph.

For $k = 0, 1, \ldots$. The expected value of K_i in state i is called the expected curtate future tenure of (x) and is denoted by $e_{x,i}$. Thus

$$e_{x,i} = \sum_{k=1}^{\infty} k \cdot \Pr(K_i = k) = \sum_{k=1}^{\infty} k \sum_{j \in \Gamma_i} {}_k p_x\,(i)\, q_{j,x+k}(i) \tag{6}$$

Observing all possible paths from the initial state through transition types of housing units (transition nodes in the graph) to the owners death (the final node), based on demographic statistics, one can calculate the expected number of housing units of different type $i, j \in H$ on the market, their supply and demand, compare it with the spatial plan in the area and therefore better manage the portfolio, avoiding the higher risk of oversupply on the housing market and therefore drop of housing prices on the market in case of oversupply. It needs a coordination with the local authorities who prepare the long-term spatial plan, also regarding the accessibility of individual housing units and dedicated land use for services. Let us denote by $H_\pi \subset H$ a set of all nodes in path π. For any such path a single or a multistage reverse mortgage equity withdrawals plan can be designed. The expected total future lifetime is

$$e_x = \sum_{i \in H_\pi} \sum_{k=1}^{\infty} k \cdot \Pr(K_i = k) = \sum_{i \in H_\pi} \sum_{k=1}^{\infty} k \sum_{j \in \Gamma_i} {}_k p_x\,(i)\, q_{j,x+k}(i) \tag{7}$$

In the next chapters we shall present how these paths (the sequences of dwelling transitions) could be better financially supported using flexible reverse mortgage schemes.

3 Multistage Reverse Mortgage Equity Withdrawals

3.1 *The Basic Loan Model of ERS with Transparent Longevity Insurance*

The loan model allows the homeowner to draw liquid amounts according to different schemes: (a) Whole amount at the closing of the contract; (b) As line of credit up to the amount stated in the contract; (c) Level amounts in the period stated in the contract.

First we need to define the minimum net property value. This is the property value at the time of death of the homeowner, less the future value of closing costs of contract and sale costs of property. With the existing loan model we charge only a portion of wealth embedded in the residential property (yearly disbursement amount and interest) each time we draw an amount of money from the value of the residential property. The resident stays the owner of the property until his or her death or until he moves to another property or to a long-term care facility. The owner always has the possibility to repay the loan, either with extra funds or by selling the property on the market. Also, the heirs have the possibility to repay the loan and keep the property at the time of owner's death.

The problem with these loan models is that the property owner's longevity is unknown, i.e., at the moment of closing the reverse mortgage contract the owners do not know whether they will live beyond the time when the wealth embedded in the residential property is exhausted (it means that the loan is equal to the net value of the property). To solve this problem we developed a longevity insurance for homeowners. In our approach we set the drawing period according to the life expectancy of the homeowner (curtailed life expectancy plus 1 year) regarding his or her age at the closing of the reverse mortgage contract. The part of the drawn amount is then used to purchase longevity insurance in a form of deferred lifetime annuity to which the homeowners are entitled after they survive the period of their expected lifetime $n = [e_x] + 1$. The reverse mortgage products facilitate the transformation of the value embedded in the residential property into the annuity stream:

$$\ddot{a}_x = \sum_{k=0}^{\infty} {}_kp_x \cdot v^k \tag{8}$$

Here v is discount factor and $\ddot{a}_x$ is the present value of future cash flow of 1 monetary unit (let us consider EUR) withdrawn from the embedded value in the housing unit at the beginning of each year for homeowner that is x years old—here homeowner (x). In the actuarial calculations we shall use the German annuity mortality tables DAV 1994R. Here ${}_kp_x$ is the probability that the homeowner who is x years old at the closing of the contract will survive the next k years in the housing unit of type $i = 0$, where v is the discounting factor. We can calculate the yearly amount of lifetime annuity which includes administrative expenses by multiplying the annuity factor with the net value of residential property. Annuity factor $fr(x, b)$ is:

$$fr(x, b) = \frac{1}{(1 + \gamma_2) \cdot \ddot{a}_x} \tag{9}$$

where γ_2 represents administrative costs associated with the disbursement of annuity that the insurance company charges for each disbursement of annuity.

In general, yearly withdrawal amount $YWA_{i=0}$ is calculated according to the net value of residential property $VP_{i=0}$ and annuity factor for (x) of gender b (male or female) $fr(x, b)$:

$$YDA_{i=0} = YWA_{i=0} - R_{i=0} = \frac{VP_{i=0}}{\ddot{a}_x \cdot (1 + \gamma_2)} - -VP_{i=0} \cdot \rho, \tag{10}$$

where the longevity insurance needs to cover yearly disbursement amount $YDA_{i=0}$ that is paid to the homeowner in the amount of interest $R_{i=0}$ on the advanced mortgage loan which depends on interest rate ρ :

$$R_{i=0} = VP_{i=0} \cdot \rho \tag{11}$$

Gross mathematical reserves V_x^a of person aged x at the end of each year in the period of pay-out the lifetime annuity are calculated:

$$V_x^a = (1 + \gamma_2) \cdot \ddot{a}_x \cdot (YDA_{i=0} + R_{i=0}) \tag{12}$$

3.2 *Flexible Multistage ERS Model*

The most important achievement in our schemes is that the longevity insurance in the form of deferred annuities stays in force also if the homeowner sells the existing property ($i = 0,\ i \in H_\pi$) and buys a new residential property $j,\ j \in H_\pi,\ j \in \Gamma_i$, of value VP_j. Here Γ_i is the set of all successors of i. In general, the homeowner can change the property many times. The value of the current property will be denoted by VP_j. We assume that $VP_j \leq VP_0$ and $R_j \leq R_0$. At each transaction we get the constant sum:

$$YDA_{i=0} + R_{i=0} = YDA_j + R_j; j \in H_\pi$$

From amount of the lifetime annuity the homeowner can cover at least part of retirement home costs if needed. In our model we are more flexible. On the horizon $n = [e_x] + 1$ explained in equation (7) the premium for longevity insurance in the form of deferred lifetime annuity $P^a_{x:\overline{n}|}$ is paid and could be calculated by (13). The percentage of administrative costs charged by an insurance company is equal to γ_1 (in the period of collecting the premiums) and γ_2 (in the period when the annuity is paid out).

$$\begin{aligned} P^a_{x:\overline{n}|} &= (YDA_{i=0} + R_{i=0}) \cdot \frac{v^n \cdot {}_np_x \cdot \ddot{a}_{x+n} \cdot (1+\gamma_2)}{\ddot{a}_{x:\overline{n}|} \cdot (1-\gamma_1)} = \\ &= (YDA_{i=0} + R_{i=0}) \cdot \frac{v^n \cdot {}_np_x \cdot \left[\sum_{s=0}^{\infty} v^s \cdot {}_sp_{x+n}\right] \cdot (1+\gamma_2)}{\left[\sum_{s=0}^{n-1} v^s \cdot {}_sp_x\right] \cdot (1-\gamma_1)} \end{aligned} \tag{13}$$

The part of the yearly withdrawn amount *YWA* on the time horizon $n = [e_x] + 1$ is used to pay the premium for the longevity insurance $P^a_{x:\overline{n|}}$. In this way, this part of the property value is transferred to mathematical reserves for longevity insurance. The premiums are accumulating and the value of mathematical reserve at the end of year k (at age of the policy holder $x + k$) ${}_kV^a_{x:\overline{n|}}$ is calculated according to the following formula:

$$\begin{aligned} {}_kV^a_{x:\overline{n|}} &= (YDA_{i=0} + R_{i=0}) \cdot v^{n-k} \cdot {}_{n-k}p_{x+k} \cdot \ddot{a}_{x+n} \cdot (1 + \gamma_2) \\ &\quad - P^a_{x:\overline{n|}} \cdot \ddot{a}_{x+k:\overline{n-k|}} \cdot (1 - \gamma_1) = \\ &= (YDA_{i=0} + R_{i=0}) \cdot v^{n-k} \cdot {}_{n-k}p_{x+k} \cdot \left[\sum_{s=0}^{\infty} v^s \cdot {}_sp_{x+n}\right] \cdot (1 + \gamma_2) \\ &\quad - P^a_{x:\overline{n|}} \cdot \left[\sum_{s=0}^{n-k-1} v^s \cdot {}_sp_{x+k}\right] \cdot (1 - \gamma_1) \end{aligned} \tag{14}$$

The yearly withdrawal amount in the period $n = [e_x] + 1$ is

$$YWA = \frac{\rho \cdot VP_{i=0}}{(1 + \rho)\left[(1 + \rho)^n - 1\right]} = \frac{R}{(1 + \rho)\left[(1 + \rho)^n - 1\right]} \tag{15}$$

The yearly withdrawal amount covers the disbursement amount, the reverse mortgage running costs, and the longevity insurance premium.

Let us denote by *ACC* the reverse mortgage account running costs and by *YDA* the yearly disbursement amount. Therefore we can write:

$$YWA = YDA + ACC + P^a_{x:\overline{n|}} \tag{16}$$

From (13) it follows:

$$P^a_{x:\overline{n|}} = (YDA + R) \cdot \frac{v^n \cdot {}_np_x \cdot \ddot{a}_{x+n} \cdot (1 + \gamma_2)}{\ddot{a}_{x:\overline{n|}} \cdot (1 - \gamma_1)} \tag{17}$$

And from (15) to (17) we can calculate the yearly disbursement amount *YDA*:

$$YDA_{i=0} = \frac{\frac{\rho \cdot VP_{i=0}}{(1+\rho)[(1+\rho)^n - 1]} - \rho \cdot VP_{i=0} \cdot \frac{v^n \cdot {}_np_x \cdot \ddot{a}_{x+n} \cdot (1+\gamma_2)}{\ddot{a}_{x:\overline{n|}} \cdot (1-\gamma_1)} - ACC}{1 + \frac{v^n \cdot {}_np_x \cdot \ddot{a}_{x+n} \cdot (1+\gamma_2)}{\ddot{a}_{x:\overline{n|}} \cdot (1-\gamma_1)}} \tag{18}$$

and $YDA_j + R_j$ is constant.

In this model we assume that the homeowner can always sell the property and buy the new residential property for the amount which is of the same value as that of the first house, or lower. Since the insurance covers the yearly disbursement amount and the interest on loan L_k in step k on horizon ($k \leq n$), the cumulative value of the loan in case the homeowner is not changing the property is

$$L_k = (1 + \rho)\frac{(1 + \rho)^k - 1}{\rho} YWA \tag{19}$$

If he/she changes the property and the value of the property changes from $VP_{i=0}$ to VP_1 the cumulative value of the loan is:

$$L_k = (1+\rho)\frac{(1+\rho)^k - 1}{\rho} YWA - VP_{i=0} + VP_1 \tag{20}$$

In general, when the homeowner sells the first property and all the following properties *j*, and buys property *j+1*, while all the properties are sold for the same amount as they were bought for, loan L_k in step *k* on horizon ($k \leq n$) would be:

$$L_k = (1+\rho)\frac{(1+\rho)^k - 1}{\rho} YWA - VP_{i=0} + VP_{j+1} \tag{21}$$

4 Numerical Example

We shall study the dynamics of the reverse mortgage income, calculated on the basis of 2 % interest rate on the outstanding loan, and 2 % discount rate embedded in the deferred lifetime annuity premium, and mortality tables for male DAV1994R at the following other parameters:

(a) Age of residential property owner at the closing of reverse mortgage contract is 65 years;
(b) Net value of property is constant, equal to 400,000 EUR;
(c) Administrative costs for payment processing and disbursement are calculated at 5 % of the premium and the annuity amount accordingly;
(d) Closing costs at each transaction are not included in the presented schemes;
(e) Yearly running costs per account are 120 EUR;
(f) Yearly premium for longevity insurance with the deferred annuity ($YDA_{i=0} + R_{i=0} = 18,879.15$ EUR) is 3,366 EUR;
(g) Yearly disbursement to homeowner until the sale of the initial property is 10,879.25 EUR (Table 1).

According to the regional statistics $e_{x,i=0} = e_{65,i=0,j=IS.} = 9$, where *IS.* means one of housing units in the retirement community, adapted for independent living (ISSC, IGLD, or ISUUS). The homeowner has a certain housing consumption in mind. He is looking for a retirement community that can take care of his needs and can provide different kinds of services like specialized catering, home cleaning, laundry, organized entertainment inside the community, and similar services. It means that he can choose one of three types of communities: (A) an independent segregated area as a special retirement village (ISSC) where the housing unit costs 160,000 EUR, or (B) an independent ground-level housing unit in a retirement community outside the city center (IGLD) where a housing unit costs 250,000 EUR, or (C) an independent living housing unit in the city center (ISUUS) where a housing unit costs 300,000 EUR.

Table 1 The initial Equity Release Scheme

Age x	Year y	Drawing amount B	B(y)+E(y−1) C	Interest amount 2.00 % D	Accumulated debt C(y)+D(y) E	Net value of contract (end of the year)
65	1	14, 365.26	14, 365.26	287.31	14, 652.56	385, 347.44
66	2	14, 365.26	29, 017.82	580.36	29, 598.17	370, 401.83
67	3	14, 365.26	43, 963.43	879.27	44, 842.70	355, 157.30
68	4	14, 365.26	59, 207.95	1, 184.16	60, 392.11	339, 607.89
69	5	14, 365.26	74, 757.36	1, 495.15	76, 252.51	323, 747.49
70	6	14, 365.26	90, 617.77	1, 812.36	92, 430.12	307, 569.88
71	7	14, 365.26	106, 795.38	2, 135.91	108, 931.28	291, 068.72
72	8	14, 365.26	123, 296.54	2, 465.93	125, 762.47	274, 237.53
73	9	14, 365.26	140, 127.73	2, 802.55	142, 930.28	257, 069.72
74	10	14, 365.26	157, 295.54	3, 145.91	160, 441.45	239, 558.55
75	11	14, 365.26	174, 806.70	3, 496.13	178, 302.84	221, 697.16
76	12	14, 365.26	192, 668.09	3, 853.36	196, 521.45	203, 478.55
77	13	14, 365.26	210, 886.71	4, 217.73	215, 104.44	184, 895.56
78	14	14, 365.26	229, 469.70	4, 589.39	234, 059.09	165, 940.91
79	15	14, 365.26	248, 424.35	4, 968.49	253, 392.83	146, 607.17
80	16	14, 365.26	267, 758.09	5, 355.16	273, 113.25	126, 886.75
81	17	14, 365.26	287, 478.50	5, 749.57	293, 228.07	106, 771.93
82	18	14, 365.26	307, 593.33	6, 151.87	313, 745.20	86, 254.80
83	19	14, 365.26	328, 110.45	6, 562.21	334, 672.66	65, 327.34
84	20	14, 365.26	349, 037.92	6, 980.76	356, 018.67	43, 981.33
85	21	14, 365.26	370, 383.93	7, 407.68	377, 791.61	22, 208.39
86	22	14, 365.26	392, 156.86	7, 843.14	400, 000.00	0.00

Therefore, in the plan we assume that when the property owner is 74 years old, his functional capacity is reduced, and therefore he decides to sell the property and to buy a new one for

1. Case A: 160,000 EUR;
2. Case B: 250,000 EUR, and
3. Case C: 300,000 EUR,

in an independent living retirement community. Tables 2, 3, and 4 show the dynamics of housing equity withdrawal where in

Case A. There is a surplus of 97,069.72 EUR after the sale of the initial residential property for 400,000 EUR and the repayment of the loan in the amount of 142,930.28 EUR. He can keep the surplus in the scheme and get 70,465.59 EUR at the end of the withdrawal period (n=22 years) or withdraw the discounted value of this amount at any time.

Table 2 The modified Equity Release Scheme after the first transaction Case A

Account x	Year y	Account A1	Interest amount 2.00 % A2	Drawing amount B	B(y)+ E(y-1) C	Interest amount 2.00 % D	Accumulated debt C(y)+D(y) E	Net value of contract (end of the year)
								257, 069.72
74	10	97, 069.72	1, 941.39	14, 365.26	14, 365.26	287.31	14, 652.56	259, 011.11
75	11	99, 011.11	1, 980.22	14, 365.26	29, 017.82	580.36	29, 598.17	231, 393.16
76	12	100, 991.34	2, 019.83	14, 365.26	43, 963.43	879.27	44, 842.70	218, 168.47
77	13	103, 011.16	2, 060.22	14, 365.26	59, 207.95	1, 184.16	60, 392.11	204, 679.28
78	14	105, 071.39	2, 101.43	14, 365.26	74, 757.36	1, 495.15	76, 252.51	190, 920.30
79	15	107, 172.81	2, 143.46	14, 365.26	90, 617.77	1, 812.36	92, 430.12	176, 886.15
80	16	109, 316.27	2, 186.33	14, 365.26	106, 795.38	2, 135.91	108, 931.28	162, 571.31
81	17	111, 502.60	2, 230.05	14, 365.26	123, 296.54	2, 465.93	125, 762.47	147, 970.18
82	18	113, 732.65	2, 274.65	14, 365.26	140, 127.73	2, 802.55	142, 930.28	133, 077.02
83	19	116, 007.30	2, 320.15	14, 365.26	157, 295.54	3, 145.91	160, 441.45	117, 886.00
84	20	118, 327.45	2, 366.55	14, 365.26	174, 806.70	3, 496.13	178, 302.84	102, 391.16
85	21	120, 694.00	2, 413.88	14, 365.26	192, 668.09	3, 853.36	196, 521.45	86, 586.42
86	22	123, 107.88	2, 462.16	14, 365.26	210, 886.71	4, 217.73	215, 104.44	70, 465.59

Table 3 The modified Equity Release Scheme after the first transaction Case B

Age x	Year y	Account A1	Interest amount 2.00% A2	Drawing amount B	B(y)+ E(y-1) C	Interest amount 2.00% D	Accumulated debt C(y)+D(y) E	Net value of contract (end of the year)
								257, 069.72
74	10	7, 069.72	141.39	14, 365.26	14, 365.26	287.31	14, 652.56	257, 211.11
75	11	7, 211.11	144.22	14, 365.26	29, 017.82	580.36	29, 598.17	227, 757.16
76	12	7, 355.34	147.11	14, 365.26	43, 963.43	879.27	44, 842.70	212, 659.75
77	13	7, 502.44	150.05	14, 365.26	59, 207.95	1, 184.16	60, 392.11	197, 260.38
78	14	7, 652.49	153.05	14, 365.26	74, 757.36	1, 495.15	76, 252.51	181, 553.03
79	15	7, 805.54	156.11	14, 365.26	90, 617.77	1, 812.36	92, 430.12	165, 531.53
80	16	7, 961.65	159.23	14, 365.26	106, 795.38	2, 135.91	108, 931.28	149, 189.60
81	17	8, 120.89	162.42	14, 365.26	123, 296.54	2, 465.93	125, 762.47	132, 520.83
82	18	8, 283.30	165.67	14, 365.26	140, 127.73	2, 802.55	142, 930.28	115, 518.69
83	19	8, 448.97	168.98	14, 365.26	157, 295.54	3, 145.91	160, 441.45	98, 176.50
84	20	8, 617.95	172.36	14, 365.26	174, 806.70	3, 496.13	178, 302.84	80, 487.47
85	21	8, 790.31	175.81	14, 365.26	192, 668.09	3, 853.36	196, 521.45	62, 444.66
86	22	8, 966.11	179.32	14, 365.26	210, 886.71	4, 217.73	215, 104.44	44, 040.99

Table 4 The modified Equity Release Scheme after the first transaction Case C

Age x	Year y	Account A1	Interest amount 2.00 % A2	Drawing amount B	B(y)+ E(y-1) C	Interest amount 2.00 % D	Accumulated debt C(y)+D(y) E	Net value of contract (end of the year)
							42, 930.28	257, 069.72
74	10	0.00	0.00	14, 365.26	57, 295.54	1, 145.91	58, 441.45	241, 558.55
75	11	0.00	0.00	14, 365.26	72, 806.70	1, 456.13	74, 262.84	225, 737.16
76	12	0.00	0.00	14, 365.26	88, 628.09	1, 772.56	90, 400.65	209, 599.35
77	13	0.00	0.00	14, 365.26	104, 765.91	2, 095.32	106, 861.23	193, 138.77
78	14	0.00	0.00	14, 365.26	121, 226.48	2, 424.53	123, 651.01	176, 348.99
79	15	0.00	0.00	14, 365.26	138, 016.27	2, 760.33	140, 776.59	159, 223.41
80	16	0.00	0.00	14, 365.26	155, 141.85	3, 102.84	158, 244.68	141, 755.32
81	17	0.00	0.00	14, 365.26	172, 609.94	3, 452.20	176, 062.14	123, 937.86
82	18	0.00	0.00	14, 365.26	190, 427.39	3, 808.55	194, 235.94	105, 764.06
83	19	0.00	0.00	14, 365.26	208, 601.19	4, 172.02	212, 773.22	87, 226.78
84	20	0.00	0.00	14, 365.26	227, 138.47	4, 542.77	231, 681.24	68, 318.76
85	21	0.00	0.00	14, 365.26	246, 046.50	4, 920.93	250, 967.43	49, 032.57
86	22	0.00	0.00	14, 365.26	265, 332.68	5, 306.65	270, 639.34	29, 360.66

Case B: There is a surplus of 7,069.72 EUR after the sale of the initial residential property for 400,000 EUR and the repayment of the loan in the amount of 142,930.28 EUR. He can keep the surplus in the scheme and get 44,040.99 EUR at the end of the withdrawal period (n=22 years) or withdraw the discounted value of this amount at any time.

Case C: There is a deficit of 42,930.28 EUR after the sale of the initial residential property for 400,000 EUR and the repayment of the loan in the amount of 100,000.00 EUR. So, the loan amount at the start of the modified scheme is 42,930.28 EUR. He still has a surplus in the amount of 29,360.66 EUR at the end of the withdrawal period (n=22 years), but cannot withdraw the discounted value of this amount before.

In the first two cases (A and B) the homeowner can buy a new property out of the proceeds of sale of the owned property after repayment of the original reverse mortgage loan, as it is required in existing reverse mortgage products which are terminated at the moment when the homeowner moves to another property. But in case C, the homeowner does not have enough money left to purchase a new housing unit according to the reverse mortgage products available up to now. The flexible reverse mortgage, as it is proposed in our paper, allows the utilization of the loan amount up to the originally contracted value when the homeowner is changing the collateral, his residential housing unit. In this case not only the transaction costs are lower but also the set of possible choices for the homeowner is larger.

At the age of 87 years after the withdrawal period $n=22$ years the equity of homeowner's residential property is exhausted completely and the longevity insurance in the form of life-time annuity starts to pay-out in the amount of 18,879.25 EUR at the beginning of each year. In case the homeowner does not change the property, then 8,000.00 EUR are used to cover interest expenses and 10,879.25 EUR is the yearly disbursement amount to the homeowner.

When the homeowner is 90 years old he needs to go to a retirement home into long-term institutional care. Therefore he sells the property in the retirement community for the same amount as was its purchasing price; he repays the loan and starts to receive higher yearly annuities in the amount of 18,879.15 EUR.

5 Conclusion

The financial crisis in Europe has caused an oversupply of housing units on the real estate market and heavy losses for older people by increasing the average retirement age, decreasing the value of pensions, decreasing the ECB interest rate, and decreasing the returns on assets held by pension funds. The stated problems present a new set of challenges regarding the management of real estate assets, especially those owned by senior citizens.

The multiple decrement model as a case of multistate transition was presented as an appropriate model for better understanding the supply and demand on the residential real estate market caused by the aging population, and can improve spatial planning based on demographic changes. The multiple decrement models allow for the calculation of the expected tenure in each type of residential property, according to the locational preferences and functional capacities of homeowners. They can also provide the theoretical bases to financial institutions which would like to develop more flexible equity release schemes that would better suit the decreasing functional capacity of their clients—senior citizens.

To better accommodate the housing needs of senior homeowners we developed a new type of reverse mortgage schemes which allow the sale of the existing home and the purchase of a new housing unit—several times if required—when transaction costs, especially longevity insurance costs, are reduced, due to the single longevity insurance contract. In this model we assumed that the homeowner can always sell the property and buy the new residential property for the amount which is of the same value as that of the first house, or lower. We described how to provide low transaction costs to enable senior homeowners to adopt their housing consumption to their housing requirements due to decreasing functional capacity and also how to increase the number of possible choices like in case C of the numerical example. The current reverse mortgage products like HECM do not allow keeping the longevity insurance, which is a novelty here. HECM and other existing reverse mortgage products require payment of closing costs each time when the homeowner wants to sell the existing property and purchase a new one. Such restrictions reduce the intensity of flows on the paths of a graph with more edges which connect a sequence of vertices from the initial to the final node on Figure 2.

While developing the flexible reverse mortgage product we have assumed that seniors typically adjust housing consumption to desired or affordable levels by moving, or by investment in home improvements. In case of low transactions costs, a senior homeowner adjusts his/her housing more often, according to desired housing consumption, usually dependent on decreased functional capacity. In the numerical examples we have shown that under the single flexible reverse mortgage contract, the homeowner has more choices in general.

Introduction of a maximum possible flexibility of such products would bring new customers to the financial institutions, which are currently deterred because the current ERS financial products seriously reduce the set of available choices in the life cycle of senior citizens. Information provided by multiple decrement models enable better management of cash flows in the asset portfolio and, moreover, enable financial institutions to prepare more attractive and flexible ERSs, tailored to senior citizens' needs.

The European Union Member States, like many other developed countries, have to cope with the demographic decline and the ageing population. How best to finance the living conditions, housing of the elderly and long-term care, have become highly topical issues in the recent years. The key issue in the financing debate has been how far people should fund their own living and extra lodging expenses and how far they should be publicly funded. To respond to this challenge, the Commission and Member States set out a number of recommendations and legal documents but not much has been done on the question of how to fund the changes of urban structures which are influenced by population aging. In order to maintain a vital society in a vital town of inhabitants, it is necessary to develop new economic and social conditions and a new kind of facility management in European urban areas. Housing needs of aged persons are satisfied if the housing is specifically designed to meet the physical, emotional, recreational, medical, and social needs of aged persons. They are changing frequently in their remaining life. The aforementioned developed models can support the way to improved planning and management of the asset portfolio in ERS products. By introducing more flexible reverse mortgage lending, the liquidity of the real estate market will increase. Simulation models based on presented approach should be utilized to examine the influence of different spatial planning and fiscal policies on the real estate market. Actuarial multiple decrement modeling will simulate demand for housing and social services, enabling new insights into spatial planning and real estate market forecasting. The mode presented does not include transaction costs, the volatility of interest rate, and the risk associated with the volatility of residential property value, which needs to be based on the theory of urban land rent. These missing ingredients will be included in the further study.

References

1. Costa-Font, J.: Housing-related well-being in older people: the impact of environmental and financial influences. Urban Stud. **50**(4), 657–673 (2013)
2. Debón, A., Montes, F., Sala, R.: Pricing reverse mortgages in Spain. Eur. Actuar. J. **3**(1), 23–43 (2013)
3. Demirkan, H., Olguntuerk, N.: A priority-based 'design for all' approach to guide home designers for independent living. Archit. Sci. Rev. **57**(2), 90–104 (2014)
4. Deshmukh, S.: Multiple Decrement Models in Insurance, An Introduction Using R. Springer, India (2012)
5. Drobne, S., Bogataj, M.: ATTREG Case study of Slovenia: The accessibility and the flow of human resources between Slovenian regions at NUTS 3 and NUTS 5 levels: The Attractiveness of European region and cities for residents and visitors. ESPON, Ljubljana (2011)
6. Duca, J.V., Kumar, A.: Financial literacy and mortgage equity withdrawals. J. Urban Econ. **80**, 62–75 (2014)
7. Gerber, H.U.: Life Insurance Mathematics, 3rd edn. Springer, Berlin, Heidelberg (1997)
8. Hui, E.C.M., Wong, F.K.W., Chung, K.W., et al.: Housing affordability, preferences and expectations of elderly with government intervention. Habitat Int. **43**, 11–21 (2014)
9. IIASA: ERD. In Online Databases (2002) Available via http://webarchive.iiasa.ac.at/Research/ERD/DB/ofsubordinatedocument.Cited15Dec2014
10. Kalache, A., Kickbusch, I.: A global strategy for healthy ageing. World Health **4**, 4–5 (1997)
11. Kalache, A., Plouffe, L., et al.: Global age friendly cities Guide. WHO Press, Geneva (2007). Available via http://www.who.int/ageing/publications.
12. Promislow, S.D.: Fundamentals of Actuarial Mathematics, 3rd edn. Wiley, Chichester (2015)
13. Quigley, J.M.: Interest-rate variations, mortgage prepayments and household mobility. Rev. Econ. Stat. **69**(4), 636–643 (1987)
14. Quigley, J.M., Vanorder, R.: Explicit tests of contingent claims models of mortgage default. J. Real Estate Financ. Econ. **11**(2), 99–117 (1995)

15. Rohe, W., van Zandt, S., McCarthy, G.: The social benefits and cost of home ownership: a critical assessment of research. Working Paper No. LIHO-01.12, Joint Center for Housing Studies, Harvard University, Harvard (2001)
16. Ronald, R., Doling, J.: Testing home ownership as the cornerstone of welfare: lessons from East Asia for the West. Hous. Stud. **27**(7), 940–961 (2012)
17. Rossi, P. H., Weber, E.: The social benefits of home ownership: empirical evidence from national surveys. Housing Policy Debate. **7**, 1–81 (1996)
18. Salaj, A. T., Jancar, J., Brus, M. S., et al.: The Development of the Real Estate Investment Fund for the Purpose of Regional Development. Lex Localis-Journal Of Local Self-Government. 9(3), 265–281 (2011)
19. Skrovankova, L., Soltes, M.: The Actuarial Models For Critical Illness Insurance. E & M Ekonomie a Manag. **12**(3), 69–76 (2009)
20. United Nations: World Urbanization Prospects. The 2014 Revision, Highlights (ST/ESA/SER.A/352). Department of Economic and Social Affairs, Population Division (2014)
21. Venti, S.F., Wise, D.A.: Moving and housing expenditure: transactions costs and disequilibrium. J. Public Econ. **23**, 207–243 (1984)

Speedup of Calibration and Pricing with SABR Models: From Equities to Interest Rates Derivatives

Ana María Ferreiro, José A. García-Rodríguez, José G. López-Salas, and Carlos Vázquez

Abstract In the more classical models for equities and interest rates evolution, constant volatility is usually assumed. However, in practice the volatilities are not constant in financial markets and different models allowing a varying local or stochastic volatility also appear in the literature. Particularly, here we consider the SABR model that has been first introduced in a paper by Hagan and coworkers, where an asymptotic closed-form formula for the implied volatility of European plain-vanilla options with short maturities is proposed. More recently, different works (Mercurio and Morini, Modeling Interest Rates: Advances in Derivatives Pricing, Risk Books 2009; Hagan and Lesniewski, LIBOR market model with SABR style stochastic volatility. Working Paper. http://lesniewski.us/papers/working/SABRLMM.pdf, 2008; Rebonato, A time-homogeneous SABR-consistent extension of the LMM. Risk, 2008) have extended the use of SABR model in the context of LIBOR market models for the evolution of forward rates (SABR-LMM). One drawback of these models in practice comes from the increase of computational cost, mainly due to the growth of model parameters to be calibrated. Additionally, sometimes either it is not always possible to compute an analytical approximation for the implied volatility or its expression results to be very complex, so that numerical methods (for example, Monte Carlo in the calibration process) have to be used. In this work we mainly review some recently proposed global optimization techniques based on Simulated Annealing (SA) algorithms and its implementation on Graphics Processing Units (GPUs) in order to highly speed up the calibration and pricing of different kinds of options and interest rate derivatives. Finally, we present some examples corresponding to real market data.

Keywords SABR volatility models • SABR/LIBOR market models • Parallel simulated annealing • GPUs

---

A.M. Ferreiro • J.A. García-Rodríguez • J.G. López-Salas • C. Vázquez (✉)
Faculty of Informatics, Department of Mathematics, Campus Elviña s/n, 15071-A Coruña, Spain
e-mail: aferreiro@udc.es; jagrodriguez@udc.es; jose.lsalas@udc.es; carlosv@udc.es

J.A. Londoño et al. (eds.), *Actuarial Sciences and Quantitative Finance*, Springer Proceedings in Mathematics & Statistics 135, DOI 10.1007/978-3-319-18239-1_4

1 Introduction

Mathematical models have become of great importance in order to price financial derivatives on different underlying assets. However, in most cases there is no explicit solution to the governing equations, so that accurate robust fast numerical methods are required. Furthermore, financial models usually depend on many parameters that need to be calibrated to market data. As in practice the valuations are required almost in real time, the speed of numerical computations becomes critical and this calibration process must be performed as fast as possible.

In the classical Black–Scholes model [1], the underlying asset follows a log-normal process with constant volatility. However, in real markets the volatilities are not constant and they can vary for each maturity and strike (volatility surface). In order to overcome this problem, different local and stochastic volatility models have been introduced. In [7], Hagan, Kumar, Lesniewski, and Woodward proposed a stochastic volatility model which is known as the SABR model (acronym for stochastic, alpha, beta, and rho, three of the four model parameters), arguing that sometimes local volatility models could not reproduce market volatility smiles and that their predicted volatility dynamics contradicts market smiles and skews. Under the asset's canonical measure, the forward price of an asset follows a CEV type process with stochastic volatility driven by a driftless process. The Brownian motion driving the volatility can be correlated with the one associated with the forward price. The main advantages of the model are the following. Firstly, it is able to correctly capture market volatility smiles. Secondly, its parameters, which play specific roles in the generation of smiles and skews, have an intuitive meaning. Thirdly, the authors obtained an analytical approximation for the implied volatility (known as Hagan formula) through singular perturbation techniques, thus allowing an easy calibration of the model. Finally, it has become the market standard for interpolating and extrapolating prices of plain vanilla options [15]. In [11] Obłój proved that for strikes far from the money and/or long maturities Hagan formula is neither arbitrage free nor a good approximation of implied volatilities. Besides, the author improved the former formula.

The existence of closed-form formula simplifies the calibration of the parameters to fit market data. However, when considering constant parameters (static SABR model), the volatility surface of a set of market data for several maturities cannot be suitably fitted. In order to cope with this problem, SABR model with time-dependent parameters (dynamic SABR) is proposed in [7]. Nevertheless, time-dependent parameters highly increase the computational cost and it is not always possible to compute an analytical approximation for the implied volatility [3]. In this case, we can use numerical methods (e.g., Monte Carlo) in the calibration process.

The standard Libor Market Model (LMM) [2] presents the same drawbacks as the classical Black–Scholes theory. The major disadvantage comes from the assumption of deterministic volatility coefficients that prevents matching cap and swaption volatility smiles and skews observed in the markets. Thus, there has been great research effort in extending the standard LMM to correctly capture market volatility smiles and skews. Several authors have recently tried to unify SABR and LIBOR market models. In the more standard LIBOR market model, the dynamics of each LIBOR forward rate under the corresponding terminal measure are assumed to be martingales with constant volatility.

In [6], Hagan et al. studied the natural extension of both the LMM and the SABR model. They used the technique of low noise expansions in order to produce accurate and workable approximations to swaption volatilities. Mercurio and Morini, arguing that a number of volatility factors lower than the number of state variables is often chosen, proposed in [9] a SABR/LIBOR market model with one single volatility factor. They designed a LIBOR market model starting from the reference SABR dynamics, with the purpose of preserving the SABR closed formula approximation. In [14], Rebonato designed a time-homogeneous SABR-consistent extension of the LMM. More precisely, the author specified financially motivated dynamics for the LMM forward rates and volatilities that match the SABR prices very close. Rebonato also suggested a simple financially justifiable and computationally affordable way to calibrate the model. For sake of brevity, in this work we only focus on the Mercurio and Morini model. Readers interested in the other two models are referred to [5].

The main objective of the present work is to efficiently calibrate plain SABR models and SABR/LIBOR market models. As computations based financial analysis should be carried out almost in real-time, an efficient robust and fast optimization algorithm has to be chosen.

In general, swaptions cannot be priced in closed form in the LMM and the main challenge of these works comes from the analytical approximations to price these derivatives. All the previous papers argue that the "brute-force" approach, which consists in calibrating the models using Monte Carlo simulation to price swaptions, is not a practical choice, because each Monte Carlo evaluation is computationally very expensive. However, in this chapter we propose the use of relatively old Simulated Annealing type algorithms [8], which becomes highly efficient when implemented using High Performance Computing techniques. This combination makes possible the calibration in a reasonable computational time.

2 SABR Model

The dynamics of the forward price and its volatility satisfy the system of stochastic differential equations

$$dF(t) = V(t)F(t)^{\beta}dW(t), \qquad F(0) = \hat{f}, \tag{1}$$

$$dV(t) = \nu V(t)dZ(t), \qquad V(0) = \alpha, \tag{2}$$

where $F(t) = S(t)e^{(r-y)(T-t)}$ denotes the *forward* price of the underlying asset S, r being the constant interest rate, and y the constant dividend yield. Moreover, $V(t)$

denotes the asset volatility process, dW and dZ are two correlated Brownian motions with constant correlation coefficient ρ (i.e., $dW(t)dZ(t) = \rho dt$), and $S(0)$ is the spot price of the asset. The parameters of the model are: $\alpha > 0$ (the volatility's reference level), $0 \leq \beta \leq 1$ (the variance elasticity), $\nu > 0$ (the volatility of the volatility), and ρ (the correlation coefficient).

This model with constant parameters is known as static SABR model. The main drawback of this static SABR model arises when market data for options with several maturities are considered. In this case, too large errors can appear in the calibration process (see, for example, [3]). In order to overcome this problem, the following dynamic SABR model allows time dependency in some parameters:

$$dF(t) = V(t)F(t)^{\beta}dW(t), \qquad F(0) = \hat{f}, \tag{3}$$

$$dV(t) = \nu(t)V(t)dZ(t), \qquad V(0) = \alpha, \tag{4}$$

where ν and the correlation coefficient ρ are time dependent, i.e., $dW(t)dZ(t) = \rho(t)dt$. As in the static SABR model, the dynamic one also provides an expression to approximate the implied volatility [13],

$$\sigma_{\text{model}}(K, \hat{f}, T) = \frac{1}{\omega}\left(1 + A_1(T)\ln(K/\hat{f}) + A_2(T)\ln^2(K/\hat{f}) + B(T)T\right), \tag{5}$$

where

$$A_1(T) = \frac{\beta - 1}{2} + \frac{\eta_1(T)\omega}{2},$$

$$A_2(T) = \frac{(1-\beta)^2}{12} + \frac{1 - \beta - \eta_1(T)\omega}{4} + \frac{4\nu_1^2(T) + 3(\eta_2^2(T) - 3\eta_1^2(T))}{24}\omega^2,$$

$$B(T) = \frac{1}{\omega^2}\left(\frac{(1-\beta)^2}{24} + \frac{\omega\beta\eta_1(T)}{4} + \frac{2\nu_2^2(T) - 3\eta_2^2(T)}{24}\omega^2\right),$$

with

$$\nu_1^2(T) = \frac{3}{T^3}\int_0^T (T-t)^2\nu^2(t)dt, \quad \nu_2^2(T) = \frac{6}{T^3}\int_0^T (T-t)t\nu^2(t)dt,$$

$$\eta_1(T) = \frac{2}{T^2}\int_0^T (T-t)\nu(t)\rho(t)dt, \quad \eta_2^2(T) = \frac{12}{T^4}\int_0^T\int_0^t\left(\int_0^s \nu(u)\rho(u)du\right)^2 dsdt. \tag{6}$$

The choice of the functions ρ and ν constitutes a very important decision. The values of $\rho(t)$ and $\nu(t)$ have to be smaller for long terms (t large) rather than for short terms (t small). In this work we consider two possibilities with exponential decay:

- *Case I*: It is more classical and corresponds to the choice

$$\rho(t) = \rho_0 e^{-at}, \quad \nu(t) = \nu_0 e^{-bt}, \tag{7}$$

with $\rho_0 \in [-1, 1]$, $\nu_0 > 0$, $a \geq 0$ and $b \geq 0$. In this case, the expressions of the functions ν_1^2, ν_2^2, η_1 and η_2^2, defined by (6), can be exactly calculated and are given by:

$$
\begin{aligned}
\nu_1^2(T) &= \frac{6\nu_0^2}{(2bT)^3}\left[\left((2bT)^2/2 - 2bT + 1\right) - e^{-2bT}\right],\\
\nu_2^2(T) &= \frac{6\nu_0^2}{(2bT)^3}\left[2(e^{-2bT} - 1) + 2bT(e^{-2bT} + 1)\right],\\
\eta_1(T) &= \frac{2\nu_0\rho_0}{T^2(a+b)^2}\left[e^{-(a+b)T} - (1 - (a+b)T)\right],\\
\eta_2^2(T) &= \frac{6\nu_0^2\rho_0^2}{T^4(a+b)^4}\left[(a+b)T(1 - e^{-2(a+b)T}) - 2(1 - e^{-(a+b)T})^2\right].
\end{aligned}
\tag{8}
$$

- *Case II*: A more general case corresponds to the choice (see [3], for details)

$$\rho(t) = (\rho_0 + q_\rho t)e^{-at} + d_\rho, \quad \nu(t) = (\nu_0 + q_\nu t)e^{-bt} + d_\nu. \tag{9}$$

  In this case, the symbolic software package Mathematica allows to calculate exactly the functions ν_1^2, ν_2^2, and η_1. However, it is not possible to obtain an explicit expression for the function η_2^2, an appropriate quadrature formula has to be used.

2.1 Calibration of the Parameters Using GPUs

The calibration of the SABR model parameters can be done using the implied volatility formula or the Monte Carlo simulation method. Usually, in trading environments the second one is not used, mainly due to its high execution times. However, if we have a parallel and efficient implementation of the Monte Carlo method, we can consider its usage in the calibration procedure.

In this work, the calibration of the parameters has been carried out with a Simulated Annealing stochastic global optimization method (see [8], for example). The algorithm consists in an external decreasing temperature loop. At each fixed temperature a Metropolis process, that can be seen as a Markov chain, is performed to compute the equilibrium state at this temperature level. This Markov chain consists of randomly choosing points in the search domain: if the value of the cost function at a new point decreases, the point is accepted; otherwise the point is randomly accepted following the Metropolis criterion, where the probability of accepting points with higher cost function value decreases with temperature. This process is repeated at each temperature level until temperature is low enough. As it is well known in the literature, SA involves a great computational cost.

In order to speed up this algorithm it must be parallelized. In [4], the authors discuss about the parallelization of the SA using GPUs. In the next Sections 2.1.1 and 2.1.2 we briefly introduce two calibration techniques which are further detailed in [3, 5].

2.1.1 Calibration with Technique I

The idea is that at each temperature level the Markov chains are distributed among the GPU threads. Among all the final reached points of the threads, the one with the minimum cost function value is selected, thus performing a reduction operation. The selected point is the starting one for all the threads in the next temperature level. The process is repeated until reaching a certain value of temperature.

The previous implementation can also be improved using multi-GPUs. In this case, the Markov chains are distributed among GPUs (e.g., if we have two GPUs, half of the chains are computed by each GPU, see Figure 1) and at each GPU the chains are distributed among the threads of this particular GPU. Before advancing to the next temperature level the best point must be computed in each GPU and then the best point of all GPUs is computed and used as starting point for all the upcoming threads of the new temperature level (see Figure 1). In order to calibrate models with many parameters, the multi-GPU version becomes more suitable, since the minimization process is very costly.

2.1.2 Calibration with Technique II

In this calibration technique the cost function is computed in GPU by a Monte Carlo method. As the Monte Carlo method is carried out inside the GPU, the SA minimization algorithm is run on CPU. In order to use all available GPUs in the system, we propose a CPU SA parallelization using OpenMP [12]. So, each OpenMP SA thread uses a GPU to assess on the Monte Carlo objective function. This approach can be easily extrapolated to a cluster of GPUs using, for example, MPI [10].

2.2 *Numerical Results*

We consider market data corresponding to the EUR/USD exchange rate. The EUR/USD spot rate is $S(0) = 1.2939$ US dollars quoted in December of 2011. In Figure 2, the whole volatility surface at maturities 3, 6, 12, and 24 months is shown. Note that the dynamic SABR model with Case II captures correctly the volatility skew. The mean relative error is 2.441714×10^{-2} and the maximum relative error is 6.954307×10^{-2}. For one GPU the speedup is around 240, while for two GPUs is near 451. More details can be found in [3].

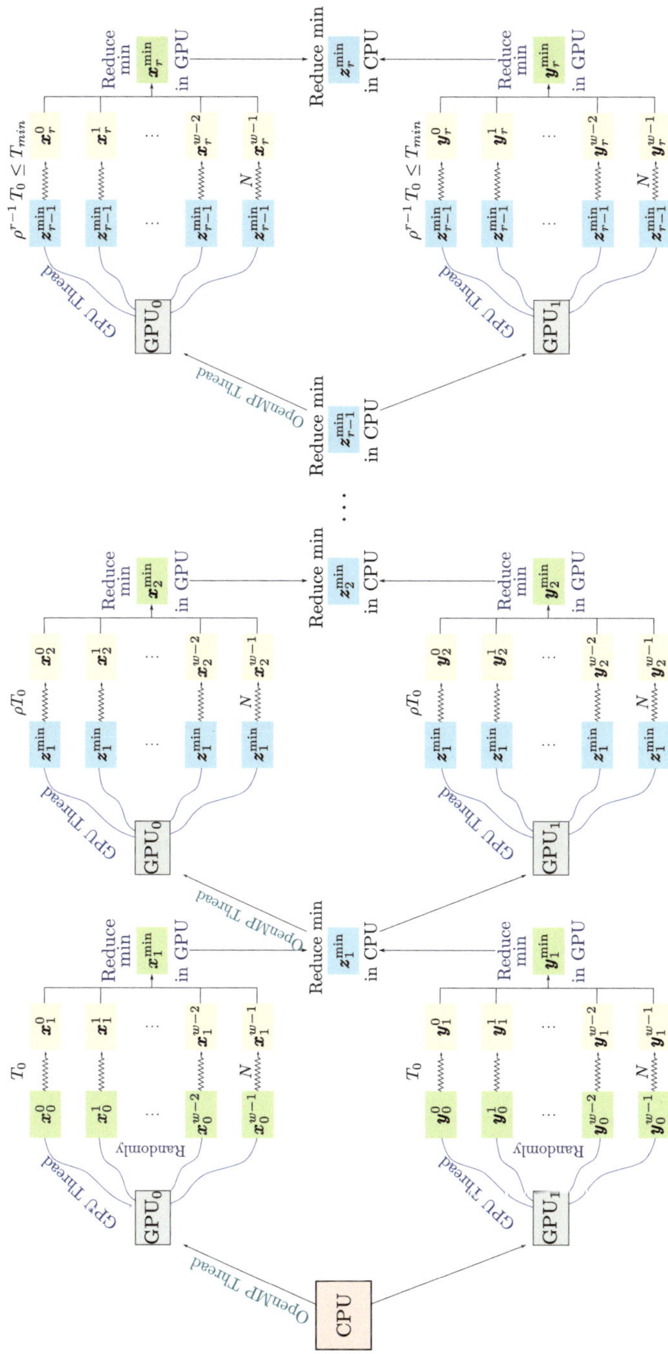

Fig. 1 Sketch of the parallel SA algorithm using two GPUs

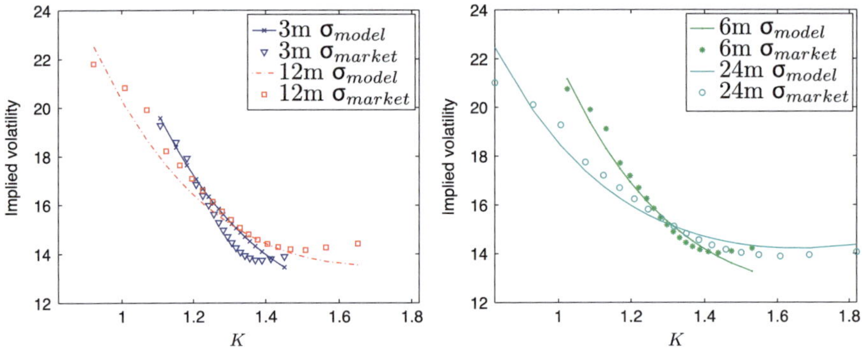

Fig. 2 EURUSD. Dynamic SABR. Case II. σ_{model} vs σ_{market} for the whole volatility surface

3 SABR/LIBOR Mercurio and Morini Model

When adding the SABR model, the forward rates and volatility processes satisfy the following coupled dynamics

$$dF_i(t) = V_i(t)F_i(t)^{\beta_i}dW_i(t),$$
$$dV_i(t) = \sigma_i V_i(t)dZ_i(t).$$

We note that if the interest rate derivative only depends on one particular forward rate, then it is convenient to use the corresponding terminal measure. However, when derivatives depend on several forward rates, a common measure needs to be used. Thus, in the case of pricing complex derivatives a change of measure produces the appearance of drift terms in forward rates and volatilities dynamics.

In the Mercurio & Morini model [9], the existence of a lognormal common volatility process to all forward rates is assumed, while each forward rate $F_i i = 1, \ldots, M$ satisfies a particular SDE. More precisely, we have

$$dF_i(t) = \mu^{F_i}(t)dt + \alpha_i V(t)F_i(t)^{\beta}dW_i(t), \tag{10}$$
$$dV(t) = \sigma V(t)dZ(t), \tag{11}$$

with the associated correlations denoted by

$$E[dW_i(t) \cdot dW_j(t)] = \rho_{i,j}dt, \quad E[dW_i(t) \cdot dZ(t)] = \phi_i dt,$$

and the initial given values $V(0) = 1$ and $F_i(0)$. Thus, the correlation structure is given by the block-matrix

$$\boldsymbol{P} = \begin{bmatrix} \boldsymbol{\rho} & \boldsymbol{\phi} \\ \boldsymbol{\phi}^\top & 1 \end{bmatrix},$$

where the submatrix $\boldsymbol{\rho} = (\rho_{i,j})$ represents the correlations between the forward rates F_i and F_j and the vector $\boldsymbol{\phi} = (\phi_1, \ldots, \phi_M)^\top$ includes the correlations between the forward rates F_i and the instantaneous volatility V.

More precisely, if we introduce the bank-account numeraire $\beta(t)$, defined by

$$\beta(t) = \prod_{j=0}^{i-1} \big(1 + \Delta t F_j(T_j)\big) \quad \text{if } t \in [T_i, T_{i+1}],$$

then, under the associated spot probability measure, the drift terms of the processes defined in (10) are

$$\mu^{F_i}(t) = \alpha_i V(t) F_i(t)^\beta \sum_{j=h(t)}^{i} \frac{\tau_j \rho_{i,j} \alpha_j V(t) F_j(t)^\beta}{1 + \tau_j F_j(t)},$$

where $h(t)$ denotes the index of the first unfixed F_i, i.e.,

$$h(t) = j, \text{ if } t \in [T_{j-1}, T_j). \tag{12}$$

The implied volatility for this model can be computed from Hagan second order approximation formula [11]:

$$\sigma\big(K, F_i(0)\big) \approx \frac{\bar{\alpha}_i}{F_i(0)^{(1-\beta)}} \times \Bigg\{ 1 - \frac{1}{2}(1 - \beta - \phi_i \sigma \omega_i) \cdot \ln\Big(\frac{K}{F_i(0)}\Big)$$
$$+ \frac{1}{12}\Big((1-\beta)^2 + (2 - 3\phi_i^2)\sigma^2 \omega_i^2 + 3\big((1-\beta) - \phi_i \sigma \omega_i\big)\Big) \cdot \Big[\ln\Big(\frac{K}{F_i(0)}\Big)\Big]^2 \Bigg\}, \tag{13}$$

where $\omega_i = \bar{\alpha}_i^{-1} F_i(0)^{(1-\beta)}$, $\bar{\alpha}_i = \alpha_i \left[e^{\int_0^{T_i} M_i(s) ds} \right]$ and $M_i(t) = -\sigma \sum_{j=h(t)}^{i} \frac{\tau_j \phi_j \alpha_j F_j(0)^\beta}{1 + \tau_j F_j(0)}$.

For the correlations, we consider the following functional parameterizations:

$$\rho_{i,j} = \eta_1 + (1 - \eta_1) \exp[-\lambda_1 |T_i - T_j|]. \tag{14}$$

3.1 Model Calibration

We consider the market prices of caplets and swaptions and we pose the corresponding global optimization problems to calibrate the model parameters. Model parameters are calibrated in two stages, firstly to caplets and secondly to swaptions. We note that model parameters can be classified into two categories (volatility and correlation parameters). The volatility parameters are $\boldsymbol{x} = (\phi_i, \sigma, \alpha_i)$ and the correlation ones $\boldsymbol{y} = (\eta_1, \lambda_1)$. According to this classification, the cost functions to be minimized in the calibration process are the following:

- Function to calibrate the market prices of caplets:

$$f_c(\boldsymbol{x}) = \sum_{i=1}^{M} \sum_{j=1}^{\text{num}K} \Big(\sigma\big(K_j, F_i(0)\big) - \sigma_{\text{market}}\big(K_j, F_i(0)\big)\Big)^2 (\boldsymbol{x}),$$

  where σ is given by Hagan formula (13), σ_{market} are the market smiles, and $\boldsymbol{x}$ is the vector containing the volatility parameters of the model. Moreover, M and $numK$ denote the number of maturities and strikes of the caplets, respectively. In order to minimize this function we use the previous calibration technique I.
- Function to calibrate the market prices of swaptions:

$$f_s(\boldsymbol{y}) = \sum_{i=1}^{\text{numSws}} \left(S_{\text{Black}}(\text{swaption}_i) - S_{\text{MC}}(\text{swaption}_i)\right)^2 (\boldsymbol{y}),$$

  where swaption$_i$ denotes the ith swaption, S_{Black}[1] is the Black formula for swaptions and $S_{\text{MC}}(\text{swaption}_\text{i})$ is the value of the ith swaption computed using Monte Carlo method.[2] Moreover, $\boldsymbol{y}$ denotes the vector containing the correlation parameters of the model and numSws is the number of swaptions. So as to optimize this function we employ the former calibration technique II.

3.2 Numerical Results

Market data correspond to the 6-month EURIBOR rate (see [5] for details). We show in Table 1 the smiles of the forward rates and in Table 2 the smiles of the swap rates.

[1] $S_{\text{Black}}(K, \sigma_{a,b}) = \text{Bl}(K, S_{a,b}(0), \sigma_{a,b}\sqrt{T_a}, 1) \sum_{i=a}^{b-1} \Delta t P(0, T_{i+1})$, where $\sigma_{a,b}$ is the volatility parameter quoted in the market, Bl is the classical Black–Scholes formula, and P denotes the discount factor.

[2] The payoff of the swaption $T_a \times (T_b - T_a)$, i.e., with maturity T_a and length of the underlying swap $T_b - T_a$, is $(S_{a,b}(T_a) - K)^+ \sum_{i=a}^{b-1} \Delta t P(T_a, T_{i+1})$, where $S_{a,b}(T_a) = \dfrac{1 - \prod_{j=a}^{b-1} \dfrac{1}{1 + \Delta t F_j(T_a)}}{\sum_{i=a}^{b-1} \Delta t \prod_{j=a}^{i} \dfrac{1}{1 + \Delta t F_j(T_a)}}$.

Table 1 Smiles of forward rates. Fixing dates (first column) and moneyness (first row)

	−80 %	−60 %	−40 %	−20 %	0 %	20 %	40 %	60 %	80 %
21-05-12	142.61 %	117.05 %	97.26 %	82.58 %	72.29 %	70.89 %	69.49 %	68.08 %	66.67 %
21-11-12	112.74 %	99.23 %	88.27 %	79.62 %	73.03 %	71.95 %	70.87 %	69.77 %	68.69 %
21-05-13	91.55 %	83.75 %	77.09 %	71.50 %	67.93 %	67.10 %	66.41 %	65.88 %	65.49 %
21-11-13	64.82 %	60.95 %	57.08 %	53.21 %	52.49 %	51.34 %	50.61 %	50.30 %	50.46 %
21-05-14	66.96 %	61.84 %	56.69 %	52.43 %	50.32 %	48.72 %	47.70 %	47.14 %	46.97 %
21-11-14	69.20 %	62.75 %	56.30 %	51.65 %	48.19 %	46.19 %	44.91 %	44.12 %	43.66 %
21-05-15	71.49 %	63.67 %	55.92 %	50.89 %	46.19 %	43.83 %	42.32 %	41.35 %	40.64 %
21-11-15	73.89 %	64.61 %	55.54 %	50.13 %	44.25 %	41.56 %	39.84 %	38.71 %	37.78 %
21-05-16	76.34 %	65.56 %	55.16 %	49.39 %	42.40 %	39.43 %	37.54 %	36.26 %	35.15 %
21-11-16	78.90 %	66.53 %	54.78 %	48.65 %	40.61 %	37.38 %	35.34 %	33.94 %	32.68 %
21-05-17	81.50 %	67.50 %	54.41 %	47.94 %	38.93 %	35.47 %	33.30 %	31.81 %	30.42 %
21-11-17	84.24 %	68.50 %	54.03 %	47.22 %	37.29 %	33.63 %	31.36 %	29.78 %	28.28 %
21-05-18	87.02 %	69.50 %	53.67 %	46.53 %	35.74 %	31.92 %	29.55 %	27.90 %	26.32 %

Table 2 Smiles of swap rates. Length of the underlying swaps (first column), maturities (second column), and moneyness (first row)

		−80%	−60%	−40%	−20%	0%	20%	40%	60%	80%
1 year	21/05/2012	122.30%	102.40%	87.12%	76.45%	70.40%	66.47%	64.20%	63.03%	62.56%
	21/11/2012	102.86%	89.97%	79.85%	72.49%	67.90%	64.58%	62.16%	60.39%	59.19%
	21/05/2013	95.64%	83.17%	73.42%	66.40%	62.10%	59.03%	56.84%	55.26%	54.18%
	21/11/2013	88.11%	76.06%	66.69%	60.00%	56.00%	53.18%	51.22%	49.84%	48.87%
2 years	21/05/2012	111.50%	91.60%	76.32%	65.65%	59.60%	55.67%	53.40%	52.23%	51.76%
	21/11/2012	89.66%	76.77%	66.65%	59.29%	54.70%	51.38%	48.96%	47.19%	45.99%
	21/05/2013	82.94%	70.47%	60.72%	53.70%	49.40%	46.33%	44.14%	42.56%	41.48%
	21/11/2013	77.81%	65.76%	56.39%	49.70%	45.70%	42.88%	40.92%	39.54%	38.57%
3 years	21/05/2012	106.40%	86.50%	71.22%	60.55%	54.50%	50.57%	48.30%	47.13%	46.66%
	21/11/2012	83.66%	70.77%	60.65%	53.29%	48.70%	45.38%	42.96%	41.19%	39.99%
	21/05/2013	78.34%	65.87%	56.12%	49.10%	44.80%	41.73%	39.54%	37.96%	36.88%
	21/11/2013	73.61%	61.56%	52.19%	45.50%	41.50%	38.68%	36.72%	35.34%	34.37%
4 years	21/05/2012	101.90%	82.00%	66.72%	56.05%	50.00%	46.07%	43.80%	42.63%	42.16%
	21/11/2012	80.26%	67.37%	57.25%	49.89%	45.30%	41.98%	39.56%	37.79%	36.59%
	21/05/2013	75.24%	62.77%	53.02%	46.00%	41.70%	38.63%	36.44%	34.86%	33.78%
	21/11/2013	70.91%	58.86%	49.49%	42.80%	38.80%	35.98%	34.02%	32.64%	31.67%
5 years	21/05/2012	96.15%	74.25%	58.83%	49.88%	47.40%	45.74%	44.61%	43.76%	43.05%
	21/11/2012	89.58%	68.82%	54.14%	45.54%	43.00%	39.36%	37.33%	36.15%	35.37%
	21/05/2013	83.91%	64.51%	50.71%	42.51%	39.90%	36.48%	34.59%	33.50%	32.76%
	21/11/2013	79.13%	61.09%	48.17%	40.37%	37.70%	34.50%	32.74%	31.75%	31.05%

3.2.1 Calibration to Caplets

The calibrated parameters are shown in Table 3. The execution time was 9.165 s: the mono-GPU SA employed 9.124 s (the cost function was evaluated roughly 112

Table 3 Mercurio & Morini model, calibration to caplets with SABR formula (13): calibrated parameters

	ϕ_i	α_i		ϕ_i	α_i
F_1	−0.7549	0.0888	F_8	−0.3661	0.0696
F_2	−0.2309	0.0842	F_9	−0.4770	0.0683
F_3	0.0666	0.0817	F_{10}	−0.5760	0.0693
F_4	0.1698	0.0662	F_{11}	−0.6615	0.0682
F_5	0.0302	0.0635	F_{12}	−0.7380	0.0682
F_6	−0.1098	0.0684	F_{13}	−0.8044	0.0669
F_7	−0.2417	0.0667			
$\sigma = 0.5986$					

Table 4 Calibration to caplets, σ_{market} vs. σ_{model}

	Smile of F_1			Smile of F_2		
Moneyness (%)	σ_{market}	σ_{model}	$\frac{\lvert\sigma_{\text{market}}-\sigma_{\text{model}}\rvert}{\sigma_{\text{market}}}$	σ_{market}	σ_{model}	$\frac{\lvert\sigma_{\text{market}}-\sigma_{\text{model}}\rvert}{\sigma_{\text{market}}}$
−40	97.26	102.19	5.07×10^{-2}	88.27	89.59	1.50×10^{-2}
−20	82.58	90.71	9.85×10^{-2}	79.62	81.81	2.75×10^{-2}
0	72.29	81.16	1.23×10^{-1}	73.03	75.77	3.74×10^{-2}
20	70.89	73.55	3.76×10^{-2}	71.95	71.47	6.69×10^{-3}
40	69.49	67.88	2.31×10^{-2}	70.87	68.91	2.77×10^{-2}
	Smile of F_3			Smile of F_4		
Moneyness (%)	σ_{market}	σ_{model}	$\frac{\lvert\sigma_{\text{market}}-\sigma_{\text{model}}\rvert}{\sigma_{\text{market}}}$	σ_{market}	σ_{model}	$\frac{\lvert\sigma_{\text{market}}-\sigma_{\text{model}}\rvert}{\sigma_{\text{market}}}$
−40	77.09	77.13	4.45×10^{-4}	57.08	55.98	1.92×10^{-2}
−20	71.50	71.99	6.92×10^{-3}	53.21	52.54	1.26×10^{-2}
0	67.93	68.27	5.11×10^{-3}	52.49	50.39	4.00×10^{-2}
20	67.10	65.96	1.69×10^{-2}	51.34	49.53	3.51×10^{-2}
40	66.41	65.07	2.03×10^{-2}	50.61	49.97	1.27×10^{-2}

million times) and the Nelder–Mead local optimization algorithm consumed the remaining time.

In Table 4, market vs. model volatilities for the smiles of F_1 to F_4 and the moneyness −40% to 40% are shown. The mean relative error considering all smiles is 3.11×10^{-2}.

Figure 3 shows the model fitting to the smiles of the first four forward rates.

3.2.2 Calibration to Swaptions

The calibrated parameters are $\eta_1 = 0.779175$ and $\lambda_1 = 2.722489$. Using two GPUs the execution time was approximately 2 h (by using a cluster of GPUs, computing time could be substantially reduced).

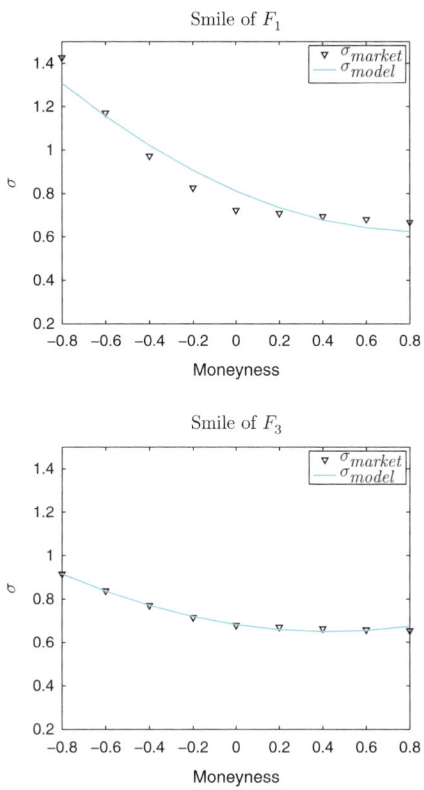

Fig. 3 σ_{market} vs. σ_{model}, smiles of $F_1, \ldots, F_4$

In Table 5, some market vs. model swaption prices are shown. The mean absolute error considering all market swaptions is 5.50×10^{-2}.

Figure 4 shows the model fitting to the first four swaption market prices.

In [5], a comparative analysis of the SABR/LIBOR models proposed by Hagan, Mercurio & Morini and Rebonato is presented. The model with the best performance is the Mercurio & Morini one, since it is the easiest to calibrate, it achieves the best fit to the swaption market prices, and it results the fastest one in the pricing with Monte Carlo simulation.

Note that the speedup with GPUs of the Monte Carlo calibration techniques can be applied to more complex products, for example CMS spread options which contain more information on the smile structure and the correlation of LIBOR rates.

Table 5 Calibration to swaptions, S_{Black} vs. S_{MC}, prices in %

	0.5 × 1 swaptions			1 × 1 swaptions		
Moneyness (%)	S_{Black}	S_{MC}	$\lvert S_{\text{Black}} - S_{\text{MC}}\rvert$	S_{Black}	S_{MC}	$\lvert S_{\text{Black}} - S_{\text{MC}}\rvert$
−40	0.4866	0.4870	4.00×10^{-4}	0.5917	0.5870	4.70×10^{-3}
−20	0.3562	0.3670	1.08×10^{-2}	0.4661	0.4699	3.80×10^{-3}
0	0.2356	0.2478	1.22×10^{-2}	0.3467	0.3517	5.00×10^{-3}
20	0.1363	0.1427	6.40×10^{-3}	0.2394	0.2422	2.80×10^{-3}
40	0.0680	0.0657	2.30×10^{-3}	0.1517	0.1514	3.00×10^{-4}
	1.5 × 1 swaptions			2 × 1 swaptions		
Moneyness (%)	S_{Black}	S_{MC}	$\lvert S_{\text{Black}} - S_{\text{MC}}\rvert$	S_{Black}	S_{MC}	$\lvert S_{\text{Black}} - S_{\text{MC}}\rvert$
−40	0.7357	0.6872	4.85×10^{-2}	0.8184	0.7465	7.19×10^{-2}
−20	0.5908	0.5516	3.92×10^{-2}	0.6603	0.5959	6.44×10^{-2}
0	0.4536	0.4170	3.66×10^{-2}	0.5118	0.4469	6.49×10^{-2}
20	0.3277	0.2951	3.26×10^{-2}	0.3754	0.3137	6.17×10^{-2}
40	0.2213	0.1957	2.56×10^{-2}	0.2587	0.2078	5.09×10^{-2}

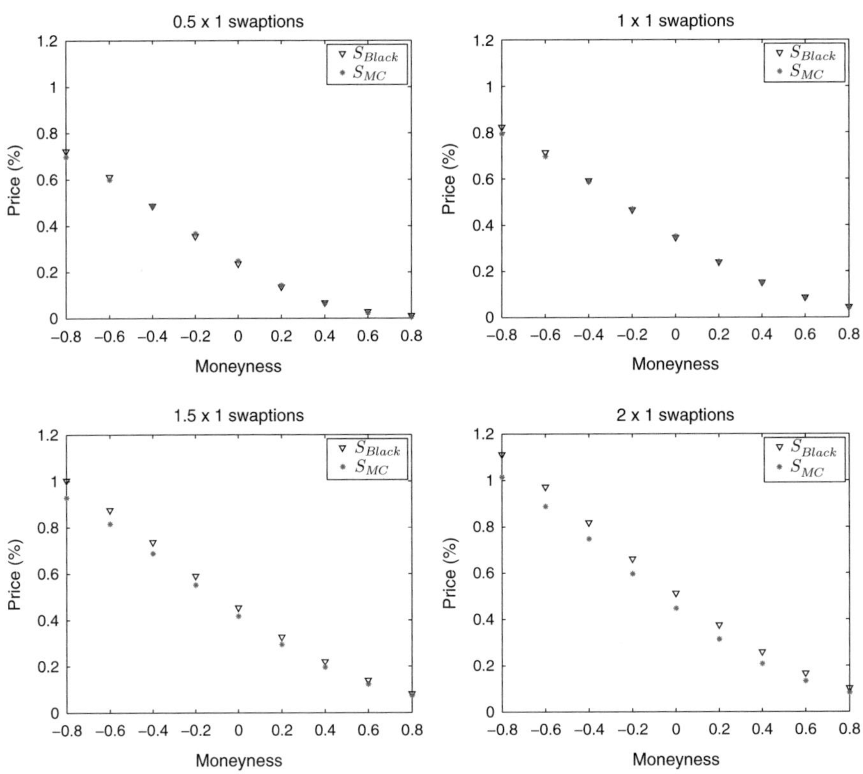

Fig. 4 S_{Black} vs. S_{MC}, $\{0.5, \ldots, 2\} \times 1$ swaptions

References

1. Black, F., Scholes, M.: The pricing of options and corporate liabilities. J. Polit. Econ. **81**, 637–654 (1973)
2. Brace, A., Gatarek, D., Musiela, M.: The Market model of interest rate dynamics. Math. Financ. **7**(2), 127–155 (1997)
3. Fernández, J.L., Ferreiro, A.M., García, J.A., López-Salas, J.G., Vázquez, C.: Static and dynamic SABR stochastic volatility models: calibration and option pricing using GPUs. Math. Comput. Simul. **94**, 55–75 (2013)
4. Ferreiro, A.M., García, J.A., López-Salas, J.G., Vázquez, C.: An efficient implementation of parallel simulated annealing algorithm in GPUs. J. Glob. Optim. **57**(3), 863–890 (2013)
5. Ferreiro, A.M., García, J.A., López-Salas, J.G., Vázquez, C.: SABR/LIBOR market models: pricing and calibration for some interest rate derivatives. Appl. Math. Comput. **242**, 65–89 (2014)
6. Hagan, P., Lesniewski, A.: LIBOR market model with SABR style stochastic volatility. Working Paper. http://lesniewski.us/papers/working/SABRLMM.pdf. (2008)
7. Hagan, P.S., Kumar, D., Lesniewski, A.S., Woodward, D.E.: Managing smile risk. Wilmott Magazine (2002)
8. Kirkpatrick, S., Gelatt, C.D., Vecchi, M.P.: Optimization by simulated annealing. Science **220**, 671–680 (1983)
9. Mercurio, F., Morini, M.: No-Arbitrage dynamics for a tractable SABR term structure Libor model. In: Modeling Interest Rates: Advances in Derivatives Pricing. Risk Books, London, UK (2009)
10. MPI: A message-passing interface standard. Message Passing Interface Forum. http://www.mpi-forum.org/docs/mpi-11-html/mpi-report.html
11. Obłój, J.: Fine-tune your smile: correction to Hagan et al. Wilmott Magazine (2008)
12. OpenMP. http://openmp.org
13. Osajima, Y.: The asymptotic expansion formula of implied volatility for dynamic SABR model and FX hybrid model. Report UTMS 2006-29, pp. 24. Graduate School of Mathematical Sciences. University of Tokyo (2006).
14. Rebonato, R.: A time-homogeneous SABR-consistent extension of the LMM. Risk **20**, 102–106 (2007)
15. Rebonato, R., Mckay, K., White, R.: The SABR/LIBOR Market Model: Pricing, Calibration and Hedging for Complex Interest-Rate Derivatives, 1st edn. Wiley, Chichester (2009)

Bergman, Piterbarg, and Beyond: Pricing Derivatives Under Collateralization and Differential Rates

Fabio Mercurio

Abstract We extend Piterbarg's result (Piterbarg, Risk, 2:97–102, 2010) on European-style derivative pricing under collateralization by relaxing the assumption of a single unsecured funding rate. Introducing different lending and borrowing rates has the effect of producing nonlinear price functionals for general claims. Buyer and seller prices diverge, and values of derivative portfolios are not the sum of the individual deal values. Conditions under which no-arbitrage price bounds can be derived explicitly are given and numerical examples showcased.

Keywords Derivatives • Collateralization • Differential rates • Funding • Nonlinear pricing • Black–Scholes • Binomial model

1 Introduction

In recent years, a large percentage of all OTC derivatives transactions has been subject to collateral agreements. Collateral agreements vary from counterparty to counterparty, and contain dealer-specific features and clauses, including collateral and currency choices. For instance, typical agreements between banks assume same-currency cash collateral and frequent posting.

Collateralization has a direct impact on the valuation of derivatives. They become similar in spirit, but not equivalent, to futures, where marking to market occurs daily and not just at the contract's maturity. Therefore, the classic risk-neutral paradigm needs to be revised to take into account collateral features and posting frequency.

Assuming cash collateral, and that collateral posting occurs in continuous time, Piterbarg derived in [21] an option pricing formula under the assumption of an unsecured funding rate that is different than the collateral rate. He showed that the price of a European-style derivative satisfies an extension of the classic Black–Scholes PDE, which can be explicitly solved, using the Feynman–Kac theorem,

F. Mercurio (✉)
Bloomberg L.P., 731 Lexington Avenue, New York, NY 10022, USA
e-mail: fmercurio@bloomberg.net

J.A. Londoño et al. (eds.), *Actuarial Sciences and Quantitative Finance*, Springer Proceedings in Mathematics & Statistics 135, DOI 10.1007/978-3-319-18239-1_5

in the extreme cases of perfect and zero collateralization. In general, Piterbarg's pricing formula shows that a funding valuation adjustment must be added to the classic risk-neutral price whenever collateralization is not perfect.

One of the crucial assumptions in [21] is that borrowing and lending rates are equal. It does not matter if the strategy replicating the option generates positive cash or requires external funding, since the rate earned when cash is positive is equal to the rate being charged when cash is negative. In short, excess cash always grows at the funding rate, regardless of its sign. This conveniently leads to a linear pricing PDE (under constant-proportion collateralization), and hence to a symmetry between buyer and seller views and associated prices.

In this chapter, we relax the assumption of a unique unsecured funding rate in Piterbarg's set-up [21], and introduce differential rates for lending and borrowing. This creates a natural asymmetry between buyer and seller prices, in some cases (namely, when no repo market for the underlying asset exists) even under full collateralization. Accordingly, derivative prices are not uniquely defined, but typically fall into no-arbitrage bands. In addition, buyer and seller prices are no longer linear functionals of derivative payoffs. For instance, the seller (resp. buyer) price of the sum of two derivatives is lower (higher) than (or equal to) the sum of their individual prices. This is because a negative excess cash generated by one deal may be partly or even totally offset by a positive cash generated by the other deal. Pricing of deals, therefore, has to be performed at portfolio level.

The derivative pricing problem under differential rates, and continuous time, was first addressed in the financial literature in [4, 14, 15]. We here extend these works by introducing collateralization, also extending the recent work [7].[1] Different buying and selling prices in the presence of differential rates have also been derived in [18]. Derivative pricing under collateralization and differential rates has also been touched upon by [1, 17]. Alternative works on credit and funding, which include differential rates, are those of [5, 9, 10, 13, 20].

We finally notice we assume no default risk. The benefits this assumption brings in terms of a simpler set-up are easy to acknowledge. But one may argue that assuming higher funding rates than the (possibly risk-free) collateral rate corresponds to assuming a nonzero credit risk, which should therefore be modeled as well, as for instance in [6] or [9], where a bank's possible default on its funding debt has also been considered (see in particular Remark 4.5 of [9] for a discussion regarding decoupling credit risk from liquidity components). However, the objective of this paper is to extend [21] in the direction of differential rates, as Bergman did in his extension [4] of the Black and Scholes model. Excluding credit also helps isolate the effect of differential rates, so as to better understand the implications of their introduction. Furthermore, even under full collateralization, *i.e.*, when credit risk has virtually zero, or very little, impact, we can derive some nontrivial results, as for instance nonlinear pricing and dependence on funding rates, which are definitely worth pointing out.

[1]This article supplements and extends the work in [19].

2 Piterbarg's Pricing Model [21]

Piterbarg introduced in [21] the problem of pricing a European-style derivative subject to collateralization, where collateral is cash posted continuously in time. The derivative, whose time-T payoff is here denoted by $H(T)$, is assumed to be written on an asset that can be sold/purchased at repo, whose real-world dynamics follows a local-volatility process:

$$dS(t) = \cdots \, dt + \sigma(t, S(t))S(t)\, dW(t) \tag{1}$$

Assuming no arbitrage and zero default risk, [21] showed that the derivative's price $V(t) = V(t, S(t))$ satisfies the following extended Black–Scholes PDE:

$$\begin{aligned} &\frac{\partial V(t)}{\partial t} + \frac{1}{2}\sigma(t,S)^2 S^2 \frac{\partial^2 V(t)}{\partial S^2} + \big(r_R(t) - r_d(t)\big)\frac{\partial V(t)}{\partial S} S \\ &= r_F(t)V(t) - \big(r_F(t) - r_c(t)\big)\mathscr{C}(t) \end{aligned} \tag{2}$$

with terminal condition $V(T) = H(T)$, where $\mathscr{C}(t)$ is the value of the collateral account, $r_c(t)$ is the collateral rate, $r_d(t)$ is the dividend yield, $r_R(t)$ is the repo rate and $r_F(t)$ is the unsecured funding rate, all at time t.

Using the Feynman–Kac formula, one then gets:

$$V(t) = E_t\Big[e^{-\int_t^T r_F(u)\,du} H(T)\Big] + E_t\Big[\int_t^T e^{-\int_t^u r_F(v)\,dv}\big(r_F(u) - r_c(u)\big)\mathscr{C}(u)\,du\Big] \tag{3}$$

where the conditional expectations are taken under the measure where the S dynamics are

$$dS(t) = \big(r_R(t) - r_d(t)\big)S(t)\,dt + \cdots$$

After rearranging terms in Eq. (2), [21] also derived an equivalent formulation that uses $r_c(t)$ as the discount rate. In general, an effective discount rate can be obtained by assuming constant-proportion collateral posting, as in [7]:

$$\mathscr{C}(t) = \gamma V(t) \tag{4}$$

where γ is a constant such that $0 \le \gamma \le 1$, and where the boundary values $\gamma = 0$ and $\gamma = 1$ correspond to the uncollateralized and fully collateralized cases, respectively. In this case, the derivative's price can be written as:

$$V(t) = E_t\Big[e^{-\int_t^T r_\gamma(u)\,du} H(T)\Big] \tag{5}$$

where $r_\gamma(t)$ is the convex linear combination of collateral and funding rates defined by $r_\gamma(t) := \gamma r_c(t) + (1-\gamma) r_F(t)$, see [7]. Therefore, under constant-proportion collateralization, the derivative's price is still equal to a risk-neutral expected discounted payoff, but with the caveat that the effective rate $r_\gamma(t)$ must be used for discounting.

We notice that, in the absence of a repo market for the underlying asset, formulas (3) and (5) remain the same, with the difference that conditional expectations must be taken under the measure where the drift rate of S is given by $r_F(t) - r_d(t)$.

Introducing the flag ψ such that $\psi = 1$ when there exists a repo market for S and $\psi = 0$ otherwise, the pricing formulas above can then be summarized, at time 0 and under constant-proportion collateralization, as follows:

$$V(0) = \mathscr{P}(H; r_F) := \tilde{E}[e^{-\int_0^T r_\gamma(t)\mathrm{d}t} H(T)] \tag{6}$$

where $\tilde{E}$ denotes expectation under the probability measure where the evolution of S is given by

$$\mathrm{d}S(t) = \big(\psi r_R(t) + (1-\psi) r_F(t) - r_d(t)\big) S(t)\,\mathrm{d}t + \cdots$$

Clearly, if $\gamma = \psi = 1$, then $\mathscr{P}(H; r_F)$ does not depend on r_F.

3 Introducing Differential Rates

We extend pricing model [21] by introducing differential rates, that is different rates for borrowing and lending. We also assume slightly more general asset price dynamics under the real-world measure P:

$$\mathrm{d}S(t) = \mu(t) S(t)\,\mathrm{d}t + \sigma(t) S(t)\,\mathrm{d}W(t)$$

where μ and σ are processes adapted to the filtration $\{\mathscr{F}_t : t \geq 0\}$ generated by the Brownian motion W.[2]

The borrowing and lending rates, which are respectively denoted by $r_b(t)$ and $r_\ell(t)$, are assumed to be deterministic and the same for all investors, with $r_b(t) \geq r_\ell(t)$ for each t. They are also assumed to be the growth rates of two corresponding bank accounts, referred to as the borrowing and lending accounts, whose unit values $B_b(t)$ and $B_\ell(t)$ evolve, respectively, as:

$$\mathrm{d}B_\ell(t) = B_\ell(t) r_\ell(t)\,\mathrm{d}t, \quad B_\ell(0) = 1$$

$$\mathrm{d}B_b(t) = B_b(t) r_b(t)\,\mathrm{d}t, \quad B_b(0) = 1$$

[2]The reason why this dynamics is more general is because $\sigma(t)$ can be path-dependent. In particular, when $\sigma(t) = \sigma(t, S(t))$ we obtain the local-volatility dynamics (1).

By definition, amounts held in the borrowing account can only be negative or null, whereas those held in the lending account can only be positive or null.

As in [21], we assume cash collateral with continuous posting. The unit value $C(t)$ of the collateral account is assumed to grow deterministically at rate $r_c(t)$:[3]

$$dC(t) = C(t)r_c(t)\,dt$$

Let us consider a general contingent claim with time-T payoff denoted, as before, by $H(T)$. Contrary to [21], however, the claim can be path-dependent. Some of the results we will state below will require the claim to be European-style. But for now, the payoff $H(T)$ can be rather general.

Extending the hedging arguments in [4, 21], we consider an investment strategy that consists of holding, at each time t, a portfolio of $\Delta(t)$ shares, $\xi(t)$ units of collateral account, $\eta_\ell(t)$ units of lending account and $\eta_b(t)$ units of borrowing account. The time-t value of the resulting portfolio is:

$$V(t) = \Delta(t)S(t) + \xi(t)C(t) + \eta_\ell(t)B_\ell(t) + \eta_b(t)B_b(t) \tag{7}$$

where, by definition of lending and borrowing accounts, $\eta_\ell(t) \geq 0$ and $\eta_b(t) \leq 0$.

We then introduce the following wealth amounts

$$\boldsymbol{\pi}(t) := \{\pi_S(t), \pi_c(t), \pi_\ell(t), \pi_b(t)\}$$

which are defined by

$$\begin{aligned}
\pi_S(t) &:= \Delta(t)S(t) \\
\pi_c(t) &:= \xi(t)C(t) \\
\pi_\ell(t) &:= \eta_\ell(t)B_\ell(t) \\
\pi_b(t) &:= \eta_b(t)B_b(t)
\end{aligned}$$

and represent the fractions of $V(t)$ being invested, respectively, in the asset, in the collateral account and in the lending and borrowing accounts. It follows from (7) that the sum of the individual wealth amounts must be equal to $V(t)$:

$$V(t) = \pi_S(t) + \pi_c(t) + \pi_\ell(t) + \pi_b(t) \tag{8}$$

The amounts $\pi_S(t)$ and $\pi_c(t)$ can be positive or negative, but the amounts invested, respectively, in the lending and borrowing accounts are such that $\pi_\ell(t) \geq 0$ and $\pi_b(t) \leq 0$. Moreover, since it is suboptimal to simultaneously hold nonzero amounts in the lending and borrowing accounts, we impose that

$$\pi_\ell(t)\pi_b(t) = 0 \tag{9}$$

[3]Note that $C(t)$ denotes the value of one collateral unit, whereas $\mathscr{C}(t)$ in (4) denotes the total collateral value. This explains the use of a different notation.

Finally, we assume that the strategy $\boldsymbol{\pi}(t)$ is self-financing, i.e.,

$$
\begin{aligned}
\mathrm{d}V(t) &= \pi_S(t)\,\frac{\mathrm{d}S(t)}{S(t)} + r_d(t)\pi_S(t)\,\mathrm{d}t + \pi_c(t)\,\frac{\mathrm{d}C(t)}{C(t)} + \pi_\ell(t)\,\frac{\mathrm{d}B_\ell(t)}{B_\ell(t)} + \pi_b(t)\,\frac{\mathrm{d}B_b(t)}{B_b(t)} \\
&= \pi_S(t)\big[\big(\mu(t) + r_d(t)\big)\,\mathrm{d}t + \sigma(t)\,\mathrm{d}W(t)\big] + \pi_c(t)r_c(t)\,\mathrm{d}t + \pi_\ell(t)r_\ell(t)\,\mathrm{d}t \\
&\quad + \pi_b(t)r_b(t)\,\mathrm{d}t \\
&= \pi_S(t)\big[\big(\mu(t) + r_d(t)\big)\,\mathrm{d}t + \sigma(t)\,\mathrm{d}W(t)\big] + \pi_c(t)r_c(t)\,\mathrm{d}t \\
&\quad + \big[V(t) - \pi_S(t) - \pi_c(t)\big]^+ r_\ell(t)\,\mathrm{d}t - \big[V(t) - \pi_S(t) - \pi_c(t)\big]^- r_b(t)\,\mathrm{d}t
\end{aligned}
\tag{10}
$$

where the last equality follows from (8) and the optimality condition (9), and where we set $x^+ := \max(x, 0)$ and $x^- := -\min(x, 0)$.

When the asset S can be purchased (resp. sold) at the repo rate r_R, it is straightforward to show that the self-financing condition becomes:

$$
\begin{aligned}
\mathrm{d}V(t) &= \pi_S(t)\big[\big(\mu(t) + r_d(t) - r_R(t)\big)\,\mathrm{d}t + \sigma(t)\,\mathrm{d}W(t)\big] + \pi_c(t)r_c(t)\,\mathrm{d}t \\
&\quad + \big[V(t) - \pi_c(t)\big]^+ r_\ell(t)\,\mathrm{d}t - \big[V(t) - \pi_c(t)\big]^- r_b(t)\,\mathrm{d}t
\end{aligned}
\tag{11}
$$

Using the flag ψ, (10) and (11) can then be summarized as:

$$
\begin{aligned}
\mathrm{d}V(t) &= \pi_S(t)\big[\big(\mu(t) + r_d(t) - \psi r_R(t)\big)\,\mathrm{d}t + \sigma(t)\,\mathrm{d}W(t)\big] + \pi_c(t)r_c(t)\,\mathrm{d}t \\
&\quad + \big[V(t) - (1-\psi)\pi_S(t) - \pi_c(t)\big]^+ r_\ell(t)\,\mathrm{d}t \\
&\quad - \big[V(t) - (1-\psi)\pi_S(t) - \pi_c(t)\big]^- r_b(t)\,\mathrm{d}t
\end{aligned}
\tag{12}
$$

A strategy $\boldsymbol{\pi}(t)$ is said to be admissible if:

- It satisfies (9) and (12);
- It replicates the final payout $H(T)$, that is $V(T) = H(T)$;
- For every rate r such that $r(t) \in [r_\ell(t), r_b(t)]$ for all $t \in [0, T]$, it satisfies

$$
E\left\{ \int_0^T [\pi_S(t)\sigma(t)]^2\,\mathrm{d}t \right\} < \infty \tag{13}
$$

under the measure where the drift rate of S is $\psi r_R(t) + (1-\psi)r(t) - r_d(t)$.[4]

There is a natural link between an economy with differential rates and economies where borrowing and lending rates are equal. This is given by the following.

[4]Under constant-proportion collateralization, $\pi_c(t) = \gamma V(t)$, and mild conditions on rates, model dynamics and payoff, the resulting backward stochastic differential equation (12) admits a unique solution. See [11] for general results.

3.1 Link to Economies with No Differential Rates

Given the amounts vector $\boldsymbol{\pi}(t) = \{\pi_S(t), \pi_c(t), \pi_\ell(t), \pi_b(t)\}$, let us define:

$$r(t) := r_\ell(t)1_{\{V(t)-(1-\psi)\pi_S(t)-\pi_c(t)\geq 0\}} + r_b(t)1_{\{V(t)-(1-\psi)\pi_S(t)-\pi_c(t)<0\}} \tag{14}$$

where $V(t)$ is the value process associated with $\boldsymbol{\pi}$. The self-financing condition (12) is then equivalent to

$$\begin{aligned} dV(t) &= \pi_S(t)\big[\big(\mu(t) + r_d(t) - \psi r_R(t)\big)\,dt + \sigma(t)\,dW(t)\big] + \pi_c(t)r_c(t)\,dt \\ &\quad + \big[V(t) - (1-\psi)\pi_S(t) - \pi_c(t)\big]r(t)\,dt \end{aligned} \tag{15}$$

Therefore, if $\boldsymbol{\pi}(t)$ is an admissible hedging strategy for the pricing problem under differential rates, then it is also an admissible hedging strategy for the pricing problem where lending and borrowing rates are the same and equal to (14).

Conversely, let $r(t)$ be any rate process that lies between the lending and borrowing rates at all times, and let $\boldsymbol{\pi}(t)$ be an admissible hedging strategy for the pricing problem where lending and borrowing rates are the same and equal to r, with $\boldsymbol{\pi}(t)$ satisfying (15). If such $r(t)$ satisfies (14), then $\boldsymbol{\pi}(t)$ is also an admissible hedging strategy for the pricing problem under differential rates.

4 The Derivative's No-Arbitrage Price Band

Let us denote by $\mathscr{A}^H$ the set of all admissible hedging strategies for the claim with payoff $H(T)$. We want to find the cheapest self-financing strategy that yields the same payoff at maturity as that of the given claim. So we want to solve the following optimization problem:

$$\text{minimize } V(0) : \boldsymbol{\pi} \in \mathscr{A}^H \tag{16}$$

Let us assume that (16) admits a unique solution, which we denote by p^H, and let us denote by $\boldsymbol{\pi}^H$ and $V^H(t)$ the associated optimal portfolio and strategy's value, respectively, where clearly $p^H = V^H(0)$. Under differential rates r_ℓ and r_b, a time-0 no-arbitrage price p of the claim $H(T)$ satisfies[5]:

$$-p^{-H} =: V^H_{\min} \leq p \leq V^H_{\max} := p^H \tag{17}$$

where p^{-H} is the minimal cost of hedging the claim $-H(T)$, which we also assume to exist and be unique.

[5] See [4, 14, 15] for results in the uncollateralized case.

The reason why (17) holds is the following. First, no investor will pay more than p^H to get $H(T)$ at time T. In fact, in case p were strictly larger than p^H, they could follow the cheaper strategy $\boldsymbol{\pi}^H$, which will grant the same payoff $H(T)$ at maturity. Second, no investor would sell the claim for less than $-p^{-H}$. In fact, they know that they can borrow $-p^{-H}$ against the same payoff $-H(T)$ by following $\boldsymbol{\pi}^{-H}$.

Using a terminology similar to that of [4, 14, 15], we will refer to $[-p^{-H}, p^H]$ as the derivative's no-arbitrage price band. With some abuse of terminology, the lower and upper bounds $-p^{-H}$ and p^H will also be referred to as the buyer price and the seller price, respectively.

Below we will consider particular cases where $-p^{-H}$ and p^H can be proven to exist and be unique. We will also provide conditions under which these price bounds can be calculated explicitly in a Black–Scholes setting. But before, we provide a general characterization results, which will also be used to prove specific properties of the price bounds.

4.1 Characterizing the Upper and Lower Bounds

Hereafter, we will assume that the proportion invested in the collateral account is, as in (4), constant and equal to some $\gamma \in [0, 1]$, that is,

$$\pi_c(t) = \gamma V(t)$$

Under constant proportional collateralization, the upper and lower bounds for the derivative's price can be characterized, respectively, as the maximum and minimum no-arbitrage prices in economies with no differential rates, over all unsecured funding rates that are comprised between the lending and borrowing rates. To express this more formally, let us pick any adapted interest rate process $r(t)$ such that:

$$r(t) \in [r_\ell(t), r_b(t)] \tag{18}$$

for every $t \in [0, T]$. In general, $r(t)$ can be a function of current and past asset prices $S(u), 0 \le u \le t$. It can be discontinuous, as in the example shown in Figure 1, but its values must be comprised between the corresponding lending and borrowing rates at all times.

Let us define the time-t excess cash associated with the strategy $\boldsymbol{\pi}^H$ by

$$\mathbf{EC}(t; \boldsymbol{\pi}^H) := (1 - \gamma)V^H(t) - (1 - \psi)\pi_S^H(t)$$

We first state the following proposition.

Proposition 1. *The uniquely defined no-arbitrage price of $H(T)$ in the economy where lending and borrowing rates are the same and equal to r must be lower than or equal to the seller's price p^H and higher than or equal to the buyer's price $-p^{-H}$. In formulas:*

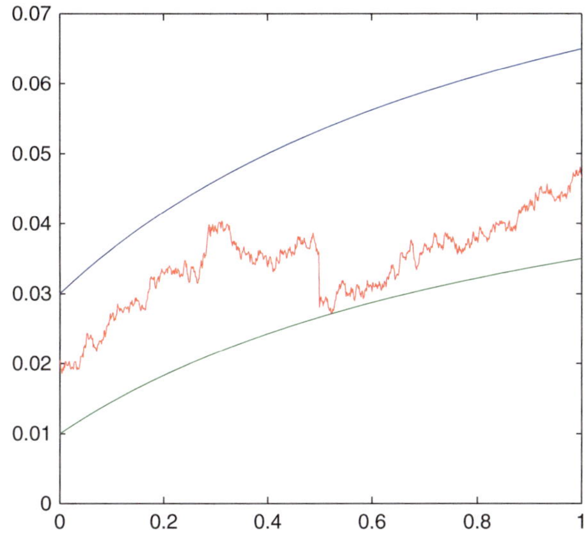

Fig. 1 An example of possible path of process r

$$-p^{-H} \leq \mathscr{P}(H;r) \leq p^{H} \tag{19}$$

where $\mathscr{P}(H;r)$ is the Piterbarg's price associated with H and r, as defined in (6).
Moreover, the upper bound p^H is attained by choosing

$$r(t) := r_\ell(t)1_{\{\mathbf{EC}(t;\pi^H)\geq 0\}} + r_b(t)1_{\{\mathbf{EC}(t;\pi^H)<0\}} \tag{20}$$

Likewise, the lower bound $-p^{-H}$ is attained by choosing:

$$r(t) := r_\ell(t)1_{\{\mathbf{EC}(t;\pi^{-H})\geq 0\}} + r_b(t)1_{\{\mathbf{EC}(t;\pi^{-H})<0\}} \tag{21}$$

Proof. *Inequalities (19) are proved in Appendix 1. Equalities (20) and (21) follow directly from Section 3.1.*

Then, combining (19) with the attainability of the upper and lower bounds, we obtain the following result, which we stated at the beginning of the section.

Corollary 1. *The seller price p^H (resp. the buyer price $-p^{-H}$) can be interpreted as the maximum (resp. minimum) of all possible no-arbitrage prices (6) corresponding to a funding rate that is smaller than the borrowing rate but larger than the lending rate at all times. In formulas:*

$$\begin{aligned} p^H &= \max\left\{\mathscr{P}(H;r):\ r_\ell(s) \leq r(s) \leq r_b(s)\ \forall s \in [0,T]\right\} \\ -p^{-H} &= \min\left\{\mathscr{P}(H;r):\ r_\ell(s) \leq r(s) \leq r_b(s)\ \forall s \in [0,T]\right\} \end{aligned} \tag{22}$$

Moreover, the price bounds are attained by using a funding rate equal to the lending rate when excess cash in the respective hedging strategy is positive and to the borrowing rate otherwise.

Note that when borrowing and lending rates are the same and equal to some r_F, that is $r_\ell \equiv r_b \equiv r_F$, then $p^H = -p^{-H} = \mathscr{P}(H; r_F)$. This constitutes an extension of Piterbarg's result [21] to path-dependent claims under the assumption of the existence and uniqueness of a self-financing strategy replicating H. Moreover, under zero collateral posting and no repo market for S, *i.e.*, when $\gamma = \psi = 0$, we recover the result of [14]. Another particular case that is worth mentioning is full collateralization under differential rates, which occurs when $\gamma = 1$ and $r_\ell \not\equiv r_b$. In this case, the possibility of buying (resp. selling) shares at repo has a direct impact on the derivative's price band. In fact, if $\psi = 1$ then the buyer and seller prices are the same, as in [21], since excess cash is zero at all times. Accordingly, payoffs are discounted at the collateral rate, and the asset's mean rate of return in the pricing measure is the repo rate (minus the dividend yield). However, if $\psi = 0$ then the buyer and seller prices are different. The corresponding excess cash is nonzero, and the pricing functional is nonlinear because of differential rates.

A final remark is on the pricing of payoffs that have the same sign in all states of the world, and are written on an asset that can be purchased/sold at repo ($\psi = 1$). In this case, p^H is obtained by discounting $H(T)$ at a weighted average of the collateral rate and the lending (resp. borrowing) rate when $H(T) \geq 0$ (resp. $H(T) \leq 0$). The opposite applies to the lower bound $-p^{-H}$.

Remark 1 (Adding Differential Repo Rates). More realistically, we may also assume differential repo rates. When lending and borrowing repo rates are different, and respectively denoted by $r_R^-(t)$ and $r_R^+(t)$, the above results remain essentially the same. In particular, the self-financing condition keeps the same form of (12), with the previously deterministic repo rate $r_R(t)$ being replaced by

$$r_R(t) = r_R^+(t)\, 1_{\{\pi_S(t)\geq 0\}} + r_R^-(t)\, 1_{\{\pi_S(t)<0\}}$$

Moreover, a characterization result on upper and lower price bounds similar to (22) still applies, but the maximum (resp. minimum) must also be taken over all repo rates that are comprised between $r_R^-(t)$ and $r_R^+(t)$ at all times t:

$$\begin{aligned} p^H &= \max\left\{\mathscr{P}(H; r, R) : r_\ell(s) \leq r(s) P \leq r_b(s),\, r_R^-(s) \leq R(s) \leq r_R^+(s) \forall s \in [0, T]\right\} \\ -p^{-H} &= \min\left\{\mathscr{P}(H; r, R) : r_\ell(s) \leq r(s) \leq r_b(s),\, r_R^-(s) \leq R(s) \leq r_R^+(s)\ \forall s \in [0, T]\right\} \end{aligned} \quad (23)$$

where price $\mathscr{P}(H; r, R)$ is calculated using a unique repo rate R.

4.2 Properties of the Upper and Lower Bounds

The pricing functional $\mathscr{P}(H;r)$ is linear in H, and hence symmetrical between buyers and sellers. Not only do buyers and sellers agree on a single derivative price, but they also price a portfolio of derivatives as the sum of the individual deal prices.

Under differential rates $r_\ell \not\equiv r_b$ and constant-proportion collateralization, $\pi_c(t) = \gamma V(t)$, the situation is more complex. Buyers and sellers do not necessarily agree on prices, and derivative prices are nonlinear functionals. In fact, we have the following.

Proposition 2. *The upper and lower bounds for a derivative's price are respectively, sublinear and superlinear functionals, in that they satisfy the following properties:*

a. *The upper and lower bounds are positively homogeneous, that is:*

$$V_{\max}^{\alpha H} = \alpha V_{\max}^{H}$$

$$V_{\min}^{\alpha H} = \alpha V_{\min}^{H}$$

for any $\alpha > 0$ and any time-T payoff $H(T)$.

b. *The upper bound $V_{\max}^{H}$ is subadditive in H, that is:*

$$V_{\max}^{H_1+H_2} \leq V_{\max}^{H_1} + V_{\max}^{H_2}$$

and the lower bound $V_{\min}^{H}$ is superadditive in H, that is:

$$V_{\min}^{H_1+H_2} \geq V_{\min}^{H_1} + V_{\min}^{H_2}$$

for any pair of time-T payoffs $(H_1(T), H_2(T))$.

Proof. *Property a follows from (22) and the linearity of $\mathscr{P}(H;r)$. To prove b, let us focus on the upper bound, and denote by r^* the optimal (upper-bound) rate for payoff $H_1(T) + H_2(T)$:*

$$r^* = \text{argmax}\big\{\mathscr{P}(H_1 + H_2; r) : r_\ell(s) \leq r(s) \leq r_b(s)\ \forall s \in [0,T]\big\}$$

so that, thanks to the linearity of $\mathscr{P}(H;r)$,

$$V_{\max}^{H_1+H_2} = \mathscr{P}(H_1 + H_2; r^*) = \mathscr{P}(H_1; r^*) + \mathscr{P}(H_2; r^*)$$

Since by Proposition 1 we have

$$V_{\max}^{H_1} \geq \mathscr{P}(H_1; r^*)$$

$$V_{\max}^{H_2} \geq \mathscr{P}(H_2; r^*)$$

we can then conclude that $V_{\max}^{H_1+H_2} \leq V_{\max}^{H_1} + V_{\max}^{H_2}$. An analogous proof holds for the lower bound.

Intuitively, property a holds because the cost of hedging α times a derivative is equal to α times the cost of hedging a unit of that derivative. Why b holds is also easy to understand. The excess cash produced by hedging a derivative with payoff $H_1(T)$ grows at different rates depending on its sign. If the excess cash is positive, it grows at the lending rate, if negative at the borrowing rate. Now add a derivative with payoff $H_2(T)$. If one hedges H_1 and H_2 separately, it may happen that one borrows money to hedge one deal (negative excess cash), at the same time investing at a lower rate the positive cash coming from hedging the other deal. Therefore, the cost of hedging the two deals jointly is lower than the cost of hedging them separately.

5 The Case of a European-Style Payoff

Proving existence and uniqueness of derivative prices p^H and $-p^{-H}$ can be hard in general unless restrictions on payoffs and asset price dynamics are introduced. To this end, let us now assume that $\sigma(t)$ is a function of time and underlying asset, $\sigma(t) = \sigma(t, S(t))$, and that the claim is European-style, so that $H(T) = h(S(T))$ for some function h, and hence $V(t) = V(t, S(t))$. Setting

$$\Delta(t) = \frac{\partial V(t)}{\partial S}$$

and using Ito's lemma and the self-financing condition (12), we get that V must satisfy the following (nonlinear) PDE:

$$\begin{aligned}&\frac{\partial V(t)}{\partial t} + \frac{1}{2}\sigma(t)^2 S^2 \frac{\partial^2 V(t)}{\partial S^2} + \big(\psi r_R(t) + (1-\psi) r_\ell(t) - r_d(t)\big) S \frac{\partial V(t)}{\partial S} \\ &\quad = r_\gamma(t) V(t) - \big(r_b(t) - r_\ell(t)\big)\left((1-\gamma)V(t) - (1-\psi) S \frac{\partial V(t)}{\partial S}\right)^-\end{aligned} \tag{24}$$

with terminal condition $V(T) = h(S(T))$, where we now set

$$r_\gamma(t) := \gamma r_c(t) + (1-\gamma) r_\ell(t).$$

We note that when $\gamma = \psi = 0$, we get the same PDE as in [4]. When the lending and borrowing rates are the same, $r_b(t) \equiv r_\ell(t)$, and $\psi = 1$, we get Piterbarg's linear PDE (2) coupled with (4).

Under mild conditions on the coefficients, see also [14], this PDE admits a unique solution, which by the Feynman–Kac theorem, see also [7], can be expressed as:

$$\begin{aligned}V(t) &= E_t\left[e^{-\int_t^T r_\gamma(u)\mathrm{d}u} h(S(T))\right] \\ &\quad + E_t\left[\int_t^T e^{-\int_t^u r_\gamma(v)\mathrm{d}v}\big(r_b(u) - r_\ell(u)\big)\left((1-\gamma)V(u) - (1-\psi)S(u)\frac{\partial V(u)}{\partial S}\right)^- \mathrm{d}u\right]\end{aligned} \tag{25}$$

where the expectation is taken with respect to the measure where the dynamics of S is:

$$dS(t) = \big(\psi r_R(t) + (1-\psi) r_\ell(t) - r_d(t)\big) S(t)\, dt + \cdots$$

Comparing (25) with (25), we see that the second expectation in its RHS represents the adjustment to the derivative's price that is induced by differential rates. Clearly, when borrowing and lending rates are equal, such term disappears and (25) reduces to (25).

Equation (25) gives an implicit representation of the derivative's price upper bound. By definition, the lower bound, as a stochastic process, is equal to the negative of the solution to the PDE (24) with terminal condition $V(T) = -h(S(T))$. Therefore, the lower bound satisfies the following PDE:

$$\begin{aligned} &\frac{\partial V(t)}{\partial t} + \frac{1}{2}\sigma(t)^2 S^2 \frac{\partial^2 V(t)}{\partial S^2} + \big(\psi r_R(t) + (1-\psi) r_\ell(t) - r_d(t)\big) S \frac{\partial V(t)}{\partial S} \\ &= r_\gamma(t) V(t) + \big(r_b(t) - r_\ell(t)\big)\left((1-\gamma)V(t) - (1-\psi) S \frac{\partial V(t)}{\partial S}\right)^+ \end{aligned} \tag{26}$$

with terminal condition $V(T) = h(S(T))$.

Since the solutions of PDE (24) and its dual PDE (26) exist and are unique, we can apply the results of Section 4 and state that the derivative price must lie in the no-arbitrage band $[V^H_{\min}(t), V^H_{\max}(t)]$, where

$$\begin{aligned} V^H_{\max}(t) &= E^{\max}_t\left[e^{-\int_t^T [\gamma r_c(u) + (1-\gamma) r_{\max}(u)]\, du} h(S(T))\right] \\ V^H_{\min}(t) &= E^{\min}_t\left[e^{-\int_t^T [\gamma r_c(u) + (1-\gamma) r_{\min}(u)]\, du} h(S(T))\right] \end{aligned} \tag{27}$$

with

$$\begin{aligned} r_{\max}(t) =& r_\ell(t) 1_{\{(1-\gamma)V^H_{\max}(t) \geq (1-\psi)\frac{\partial V^H_{\max}(t)}{\partial S} S(t)\}} + r_b(t) 1_{\{(1-\gamma)V^H_{\max}(t) < (1-\psi)\frac{\partial V^H_{\max}(t)}{\partial S} S(t)\}} \\ r_{\min}(t) =& r_\ell(t) 1_{\{(1-\gamma)V^H_{\min}(t) \leq (1-\psi)\frac{\partial V^H_{\min}(t)}{\partial S} S(t)\}} + r_b(t) 1_{\{(1-\gamma)V^H_{\min}(t) > (1-\psi)\frac{\partial V^H_{\min}(t)}{\partial S} S(t)\}} \end{aligned} \tag{28}$$

and where the expectations are taken under the measures where the mean rates of return of S are, respectively, $\psi r_R(t) + (1-\psi) r_{\max}(t) - r_d(t)$ and $\psi r_R(t) + (1-\psi) r_{\min}(t) - r_d(t)$.[6]

In the case of European-style payoffs, we can thus prove existence and uniqueness of the upper and lower price bounds. Equation (27) shows that these bounds can be represented as expectations of the derivative's payoff discounted at corresponding effective rates. However, formulas (27) are only implicit since the pricing measures

[6]We stress we still need to assume that the stochastic integral we get in Appendix 1 is a martingale.

and discount rates on its RHS depend on the derivative's prices on its LHS. In the following section, we will provide conditions under which the upper and lower bounds can be calculated explicitly.

Remark 2 (Stochastic-Control Problems). Equation (22) implies that prices $V^H_{\max}$ and $V^H_{\min}$ solve corresponding stochastic-control problems where the stochastic control is the funding rate r. Therefore, $V^H_{\max}$ and $V^H_{\min}$ must respectively satisfy the following HJB equations:

$$\begin{aligned}
&\frac{\partial V^H_{\max}(t)}{\partial t} + \max_{z\in\{\ell,b\}} \Bigg\{ \big(\psi r_R(t) + (1-\psi)r_z(t) - r_d(t)\big)\frac{\partial V^H_{\max}(t)}{\partial S} S \\
&\qquad + \frac{1}{2}\sigma(t,S)^2 S^2 \frac{\partial^2 V^H_{\max}(t)}{\partial S^2} - [\gamma r_c(t) + (1-\gamma)r_z(t)]V^H_{\max}(t) \Bigg\} = 0 \\
&\frac{\partial V^H_{\min}(t)}{\partial t} + \min_{z\in\{\ell,b\}} \Bigg\{ \big(\psi r_R(t) + (1-\psi)r_z(t) - r_d(t)\big)\frac{\partial V^H_{\min}(t)}{\partial S} S \\
&\qquad + \frac{1}{2}\sigma(t,S)^2 S^2 \frac{\partial^2 V^H_{\min}(t)}{\partial S^2} - [\gamma r_c(t) + (1-\gamma)r_z(t)]V^H_{\min}(t) \Bigg\} = 0
\end{aligned} \tag{29}$$

where we used the fact that the expressions between curly brackets are linear in $r_z(t)$, so that maximum and minimum can only be attained at the boundary of the possible values of $r_z(t)$.[7]

Remark 3 (Uncertain Funding Rate). Avellaneda et al. in [2] considered the derivative pricing problem under a stochastic volatility whose dynamics is unknown but whose values are within a given range at all times. A similar pricing problem can be considered by assuming a unique funding rate that evolves stochastically in the range set by the lending and borrowing rates. In this case, the derivative's price bounds can be shown to satisfy the same stochastic-control problems as the above price bounds $V^H_{\max}$ and $V^H_{\min}$. Therefore, we can conclude that prices $V^H_{\max}$ and $V^H_{\min}$ coincide with the derivative's price bounds in the uncertain rate model where the (unique) funding rate $r(t)$ is stochastic and such that $r(t) \in [r_\ell(t), r_b(t)]$ for each $t \in [0, T]$.

6 Payoffs Admitting Explicit Price Bounds

The explicit calculation of a derivative's upper and lower price bounds is not possible in general. However, there are specific European-style payoffs that allow for closed-form formulas. To show this, we assume that the asset price $S(t)$ follows Black–Scholes dynamics with constant volatility σ, that is $\sigma(t) \equiv \sigma$.

[7] Stochastic control problems where the optimal control can only take two values go back to Beneš's "Bang-Bang" control problem, see [3].

Both [4] and [15] showed that if excess cash at maturity is always positive (resp. negative) in almost all states of the world, then excess cash is always positive (resp. negative) at previous times as well. This allows us to remove the nonlinearity in the pricing PDE (24), which then becomes easy to solve. A similar result can also be obtained under nonzero collateralization. In fact, let us define the following deterministic rates

$$r_{\gamma,z}(t) := \gamma r_c(t) + (1-\gamma) r_z(t), \quad z \in \{\ell, b\}$$

and denote by E_t^z the conditional expectation with respect to the measure where the evolution of S is given by

$$\mathrm{d}S(t) = \big(\psi r_R(t) + (1-\psi) r_z(t) - r_d(t)\big) S(t)\,\mathrm{d}t + \cdots \quad , z \in \{\ell, b\}$$

Let us denote by $h'(x)$ the derivative of h at point x. Adapting [15] proof to the nonzero γ case, we obtain the following.

Proposition 3. *If the function h satisfies*

$$(1-\gamma)h(x) - (1-\psi)h'(x)x \geq 0 \tag{30}$$

for (almost) every x, then

$$V_{\max}(t) = E_t^{\ell}\Big[e^{-\int_t^T r_{\gamma,\ell}(u)\,\mathrm{d}u}\, h(S(T))\Big]$$

$$V_{\min}(t) = E_t^{b}\Big[e^{-\int_t^T r_{\gamma,b}(u)\,\mathrm{d}u}\, h(S(T))\Big]$$

Likewise, if the function h satisfies

$$(1-\gamma)h(x) - (1-\psi)h'(x)x \leq 0 \tag{31}$$

for (almost) every x, then

$$V_{\max}(t) = E_t^{b}\Big[e^{-\int_t^T r_{\gamma,b}(u)\,\mathrm{d}u}\, h(S(T))\Big]$$

$$V_{\min}(t) = E_t^{\ell}\Big[e^{-\int_t^T r_{\gamma,\ell}(u)\,\mathrm{d}u}\, h(S(T))\Big]$$

Proof. *Let us pick a deterministic rate $r(t)$ that satisfies (18), and set*

$$V(t, S(t); r) := E_t\Big[e^{-\int_t^T [\gamma r_c(u) + (1-\gamma) r(u)]\,\mathrm{d}u}\, h(S(T))\Big]$$

under the measure where the drift rate of S is $r_S(t) := \psi r_R(t) + (1-\psi) r(t) - r_d(t)$. Since S follows Black–Scholes dynamics with constant volatility σ, we can write

$$V(t, S; r) = e^{-\int_t^T [\gamma r_c(u) + (1-\gamma) r(u)]\,\mathrm{d}u} \int_{-\infty}^{+\infty} h(x(z)) \frac{1}{\sqrt{2\pi}} e^{-\frac{1}{2}z^2}\,\mathrm{d}z$$

where

$$x(z) := S\,e^{\int_t^T r_S(u)\,\mathrm{d}u - \frac{1}{2}\sigma^2(T-t) + \sigma\sqrt{T-t}\,z}$$

Then

$$(1-\gamma)V(t,S;r) - (1-\psi)\frac{\partial V(t,S;r)}{\partial S}S$$
$$= e^{-\int_t^T [\gamma r_c(u)+(1-\gamma)r(u)]\,\mathrm{d}u} \int_{-\infty}^{+\infty} [(1-\gamma)h(x(z)) - (1-\psi)x(z)h'(x(z))]\frac{1}{\sqrt{2\pi}}\,e^{-\frac{1}{2}z^2}\,\mathrm{d}z$$

has clearly the same sign as $(1-\gamma)h(x) - (1-\psi)h'(x)x$.

Finally, we can verify that $V(t,S;r_\ell)$ *and* $V(t,S;r_b)$ *satisfy, respectively, (24) and (26) when (30) holds. The reverse applies when (31) holds, which completes the proof.*

Conditions (30) or (31) are satisfied by European puts or calls and, more generally, by European-style derivatives whose associated excess cash has a constant sign throughout the life of the claim. The examples of calls, puts and forwards are explicitly analyzed in the following section.

We also remark that the price bounds $V_{\max}(t)$ and $V_{\min}(t)$ are linear on portfolios of claims all satisfying (30) or all satisfying (31). Moreover, if $\gamma = \psi = 1$, we recover the previously mentioned result that, despite of differential rates, buyer and seller prices are the same and only dependent on the collateral and repo rates.

When also assuming differential repo rates as in Remark 1, conditions (30) or (31) must be supplemented with a condition on the sign of h'. Therefore, when $\psi = 1$, we effectively have four different cases. In the resulting pricing formulas, the repo rate $r_R(t)$ must be replaced by $r_R^+(t)$ when $h'(x) \geq 0$ for almost all x, and by $r_R^-(t)$ when $h'(x) < 0$ for almost all x.

6.1 Particular Cases

Let us assume constant rates for simplicity, and let us focus on the $\psi = 0$ case of no repo market for S. A European call, for which $h(x) = (x-K)^+$, satisfies condition (31). Straightforward algebra then shows that the European call price bounds are given by:

$$V_{\max}(t) = e^{\gamma(r_b - r_c)(T-t)}\,\mathrm{BScall}(t,T,K,\sigma,r_b)$$
$$V_{\min}(t) = e^{\gamma(r_\ell - r_c)(T-t)}\,\mathrm{BScall}(t,T,K,\sigma,r_\ell)$$

where $\mathrm{BScall}(t,T,K,\sigma,r)$ (resp. BSput) denotes the time-t Black–Scholes call (resp. put) option price when the interest rate is r. Instead, a European put, for which $h(x) = (K-x)^+$, satisfies (30). Hence, the European put price bounds are given by:

$$V_{\max}(t) = e^{\gamma(r_\ell - r_c)(T-t)}\,\mathrm{BSput}(t,T,K,\sigma,r_\ell)$$
$$V_{\min}(t) = e^{\gamma(r_b - r_c)(T-t)}\,\mathrm{BSput}(t,T,K,\sigma,r_b)$$

In both the call and put cases, the larger the difference between r_b and r_ℓ, the wider the spread between upper and lower bounds. Numerical examples of upper and lower bounds $V_{\max}(0)$ and $V_{\min}(0)$ of call and put prices for different values of the collateralization constant γ are shown in Figure 2.

Finally, a forward contract, for which $h(x) = x - K$, satisfies condition (31). Its upper and lower price bounds are then given by

$$V_{\max}(t) = S(t)\, e^{[\gamma(r_b - r_c) - r_d](T-t)} - K\, e^{-r_{\gamma,b}(T-t)}$$

$$V_{\min}(t) = S(t)\, e^{[\gamma(r_\ell - r_c) - r_d](T-t)} - K\, e^{-r_{\gamma,\ell}(T-t)}$$

We remark that general payoffs, and portfolios of vanillas in particular, do not satisfy either (30) or (31). Simple, yet important, examples are straddles and butterflies. For these derivatives, one has either to numerically solve the associated PDEs (24) and (26), see for instance [12] or to implement binomial trees approximating a geometric Brownian motion, see Appendix 2. In Figures 3 and 4, we showcase examples of straddle and butterfly price bounds, obtained by using a binomial tree, and compare them with the values obtained by summing up the bounds for the constituent vanilla options. It is evident, especially in the butterfly case, that summing up the explicit price formulas for the underlying options can lead to large pricing errors for the whole payoff. This is because there are major cancellation effects between long and short positions, particularly for the butterfly, when we regard those derivatives as option portfolios.

7 Conclusions

We have extended Piterbarg's derivative pricing approach [21] by introducing different rates for lending and borrowing. Under differential rates, derivative prices fall into a no-arbitrage band, which is defined by upper and lower price bounds. The upper and lower bounds can be characterized, respectively, as maximum and minimum no-arbitrage prices over all non-differential funding rates that are comprised between the lending and borrowing rates. Accordingly, the upper and lower bounds of derivative portfolios are not equal to the sum of the constituent deal bounds. In fact, the upper bound is subadditive. The lower bound is superadditive.

The upper and lower bounds can be explicitly calculated in the Black–Scholes model for specific payoffs, and in particular for European calls and puts and for forward contracts. More general European-style derivatives, such as for instance straddles or butterflies, can be priced numerically by solving related nonlinear PDEs or using binomial trees.

We have also touched upon differential repo rates, whose introduction neither substantially alters our results nor compromises analytical tractability.

Finally, one could consider different collateralization rules. For instance, explicit price bounds in a Black–Scholes set-up can also be derived for positive

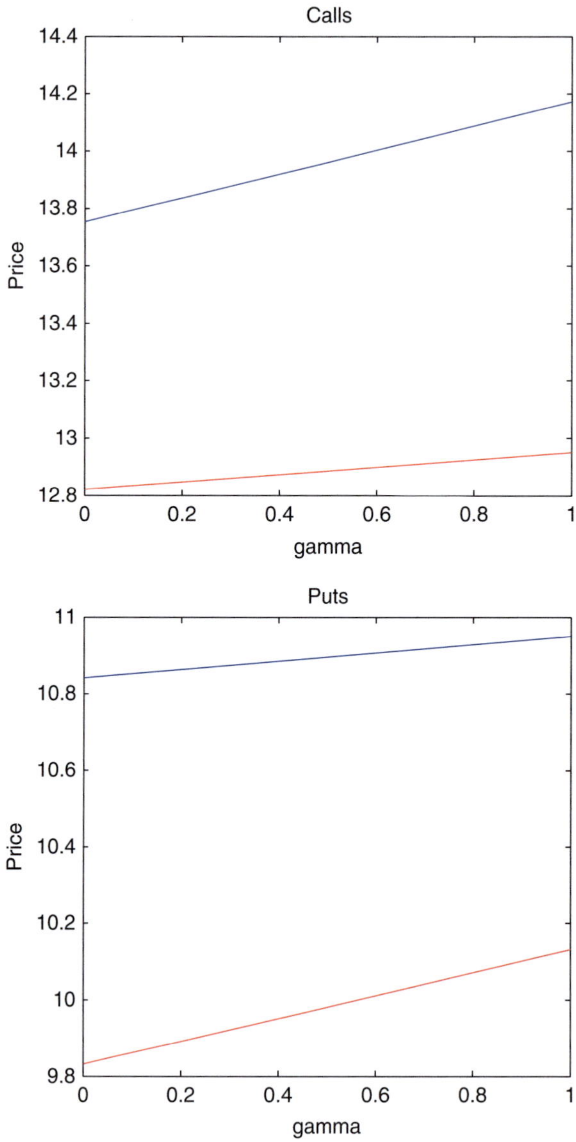

Fig. 2 Upper and lower price bounds for calls (*left*) and puts (*right*) as functions of γ. We set: $\psi = 0$, $\sigma = 0.3$, $S_0 = 100$, $K = 100$, $T = 1$, $r_d = 0$, $r_c = 0.01$, $r_\ell = 0.02$, $r_b = 0.04$

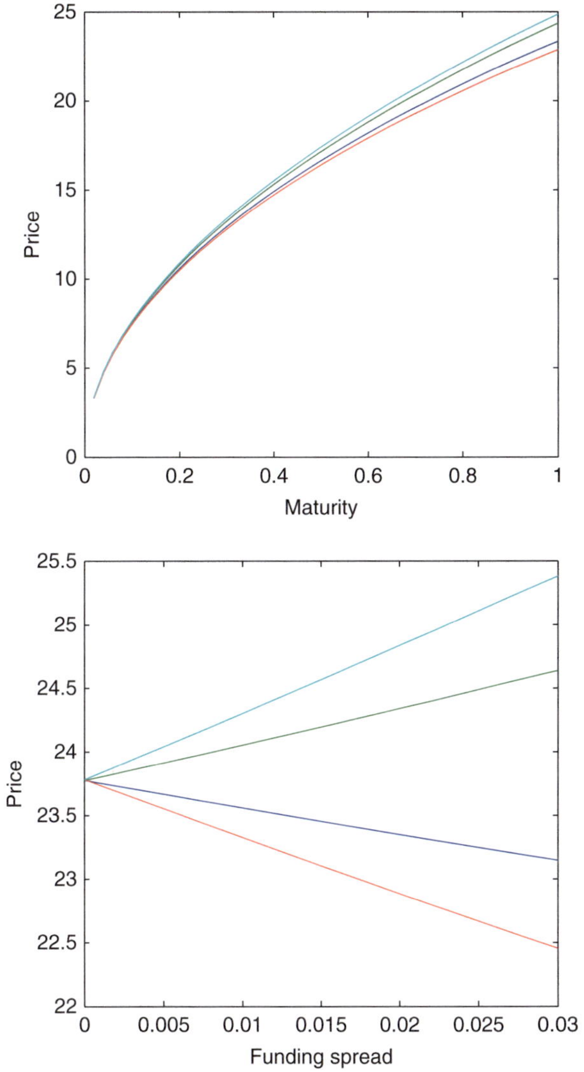

Fig. 3 Upper and lower price bounds for straddles compared with the sum of individual bounds. We set: $\psi = 0$, $\gamma = 0.5$, $\sigma = 0.3$, $S_0 = 100$, $K = 100$, $T = 1$ (*right figure*), $r_d = 0$, $r_c = 0.01$, $r_\ell = 0.02$, $r_b = 0.04$ (*left figure*). Funding spread is the difference between borrowing and lending rates

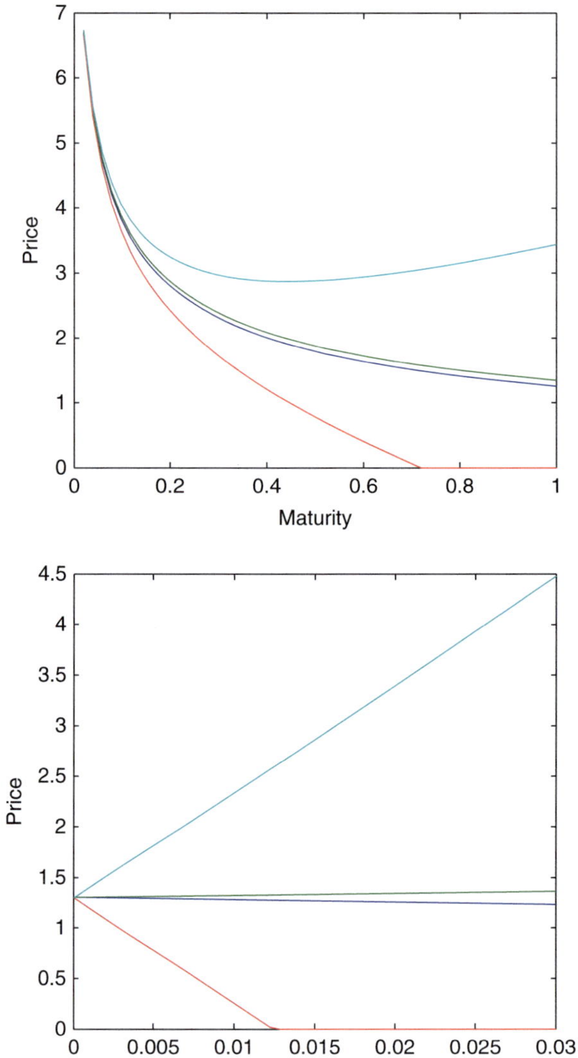

Fig. 4 Upper and lower price bounds for butterflies compared with the sum of individual bounds. We set: $\psi = 0$, $\gamma = 0.5$, $\sigma = 0.3$, $S_0 = 100$, $K_1 = 90$, $K_2 = 100$, $K_3 = 110$, $T = 1$ (*right figure*), $r_d = 0$, $r_c = 0.01$, $r_\ell = 0.02$, $r_b = 0.04$ (*left figure*) Again, funding spread is the difference between borrowing and lending rates

(resp. negative) payoffs in case of asymmetric CSA, that is when $\gamma = 1$ for $V(t) < 0$ (or $V(t) > 0$) and zero otherwise. Further extensions could be studied by introducing stochastic interest rates, multiple assets, or collateral in different currencies.

Acknowledgements Stimulating discussions with Tom Bielecki, Peter Carr, Antonio Castagna, Bruno Dupire, Julien Guyon, Minqiang Li, and Weiwei Shen are gratefully acknowledged. Needless to say, all errors are the author's responsibility. The contents of this chapter represent the authors' views only, and do not represent the opinions of any firm or institution.

Appendix 1: Proof of Proposition 1

We here prove inequalities (19) under the assumption there exist unique admissible strategies replicating $H(T)$ and $-H(T)$ respectively. We assume $\psi = 0$ for ease of notation. The case $\psi = 1$ is analogous and omitted for brevity.

Let us define the rate process $\tilde{r}$

$$\tilde{r}(t) := \gamma r_c(t) + (1-\gamma)r(t) \tag{32}$$

and the associated bank account $\tilde{B}(t) := e^{\int_0^t \tilde{r}(u)\mathrm{d}u}$. Let $\boldsymbol{\pi} = \boldsymbol{\pi}^H$ be the optimal hedging strategy for the pricing problem (16), and define the associated normalized value process $\tilde{V}(t) := V(t)/\tilde{B}(t)$. Then, we have:

$$\begin{aligned}
\mathrm{d}\tilde{V}(t) = \frac{1}{\tilde{B}(t)}\Big\{&\pi_S(t)\big[\big(\mu(t)+r_d(t)\big)\,\mathrm{d}t + \sigma(t)\,\mathrm{d}W(t)\big] + \gamma V(t) r_c(t)\,\mathrm{d}t\\
&+ \big[(1-\gamma)V(t) - \pi_S(t)\big]^+ r_\ell(t)\,\mathrm{d}t - \big[(1-\gamma)V(t)-\pi_S(t)\big]^- r_b(t)\,\mathrm{d}t\\
&- V(t)\tilde{r}(t)\,\mathrm{d}t\Big\}\\
\mathrm{d}\tilde{V}(t) = \frac{1}{\tilde{B}(t)}\Big\{&\pi_S(t)\big[\big(\mu(t)+r_d(t)-r(t)\big)\,\mathrm{d}t + \sigma(t)\,\mathrm{d}W(t)\big]\\
&+ \big[(1-\gamma)V(t) - \pi_S(t)\big]^+ r_\ell(t)\,\mathrm{d}t - \big[(1-\gamma)V(t)-\pi_S(t)\big]^- r_b(t)\,\mathrm{d}t\\
&- \big[(1-\gamma)V(t)-\pi_S(t)\big]r(t)\,\mathrm{d}t\Big\}
\end{aligned}$$

which, in integral form, becomes:

$$\begin{aligned}
\tilde{V}(t) = V(0) + \int_0^t \frac{\pi_S(u)}{\tilde{B}(u)}\sigma(u)\,\mathrm{d}\tilde{W}(u) + \int_0^t \frac{1}{\tilde{B}(u)}\Big\{&\big[(1-\gamma)V(u)-\pi_S(u)\big]^+ r_\ell(u)\\
- \big[(1-\gamma)V(u)-\pi_S(u)\big]^- r_b(u) - \big[(1-\gamma)V(u)-\pi_S(u)\big]r(u)\Big\}\,\mathrm{d}u&
\end{aligned}$$

where

$$\mathrm{d}\tilde{W}(t) := \tilde{\vartheta}(t)\,\mathrm{d}t + \mathrm{d}W(t)$$
$$\tilde{\vartheta}(t) := \frac{\mu(t) + r_d(t) - r(t)}{\sigma(t)} \tag{33}$$

Since, by definition of $r(t)$,

$$\begin{aligned}&\big[(1-\gamma)V(u) - \pi_S(u)\big]^+ r_\ell(u) - \big[(1-\gamma)V(u) - \pi_S(u)\big]^- r_b(u)\\ &\leq \big[(1-\gamma)V(u) - \pi_S(u)\big] r(u),\end{aligned}$$

for all u, the last integral term is negative, and hence

$$\tilde{V}(t) \leq V(0) + \int_0^t \frac{\pi_S(u)}{\tilde{B}(u)}\sigma(u)\,\mathrm{d}\tilde{W}(u)$$

for each t.

The stochastic integral on the RHS is a martingale, and not just a local martingale, because of (13) and the boundedness of $1/\tilde{B}$. Then,

$$\tilde{E}[\tilde{V}(T)] \leq \tilde{E}\left\{V(0) + \int_0^T \frac{\pi_S(u)}{\tilde{B}(u)}\sigma(u)\,\mathrm{d}\tilde{W}(u)\right\} = V(0)$$

where $\tilde{E}$ denotes expectation under the probability measure $\tilde{P}$ defined by the Radon–Nikodym derivative:

$$\frac{\mathrm{d}\tilde{P}}{\mathrm{d}P} = \exp\left\{-\frac{1}{2}\int_0^T \tilde{\vartheta}^2(t)\,\mathrm{d}t - \int_0^T \tilde{\vartheta}(t)\,\mathrm{d}W(t)\right\}$$

Hence,

$$\tilde{E}\Big[e^{-\int_0^T \tilde{r}(t)\mathrm{d}t} H(T)\Big] \leq V(0) = p^H \tag{34}$$

Likewise, let $\boldsymbol{\pi} = \boldsymbol{\pi}^{-H}$ be the optimal hedging strategy for the pricing problem (16) with final payoff $-H(T)$. Applying the same procedure we just followed for $\boldsymbol{\pi}^H$, we get

$$\tilde{E}\Big[e^{-\int_0^T \tilde{r}(t)\mathrm{d}t}[-H(T)]\Big] \leq p^{-H}$$

and hence

$$\tilde{E}\Big[e^{-\int_0^T \tilde{r}(t)\mathrm{d}t} H(T)\Big] \geq= -p^{-H} \tag{35}$$

Appendix 2: Pricing with a Binomial Model

The binomial option pricing model was introduced by [8, 22]. We adhere to the CRR formulation [8] and consider a set of equi-spaced discrete times $0 =: t_0 < t_1 < \ldots < t_N =: T$, denoting by S_k the asset price at time t_k. The CRR binomial model is iteratively defined as follows:

$$\frac{S_k}{S_{k-1}} = \begin{cases} u & \text{with probability } p \\ d & \text{with probability } 1-p \end{cases} \qquad k = 1, \ldots, N \tag{36}$$

where S_0 is the time-0 asset's value, and u, d, and p are such that $u > 1$, $0 < d < 1$, and $0 < p < 1$.

We extend the CRR pricing model by introducing collateralization and differential rates. The borrowing and lending rates, which are assumed to be deterministic and the same for all investors, are the growth rates of two corresponding bank accounts, referred to as the borrowing and lending accounts. Let us denote by R_b and R_ℓ the constant one-period returns of the borrowing and lending accounts, respectively.

Let us assume that $0 < d < R_\ell \leq R_b < u$ to avoid arbitrage and ill-posed cases. Let us also assume, as in CRR, that the asset cannot be purchased or sold at repo ($\psi = 1$), and that dividends are zero to lighten the notation. Then, extending the CRR hedging argument, and fixing a node in the tree, the value of the replicating strategy on that node is

$$V = \Delta S + B_\ell - B_b + C \tag{37}$$

where V denotes the strategy's value, C the collateral's value, B_ℓ the dollar amount invested in the lending account and B_b the (negative of the) dollar amount invested in the borrowing account, all on the given node, where we dropped time subscripts for ease of notation. By definition, $B_b, B_\ell \geq 0$ and $B_b B_\ell = 0$, because holding simultaneously nonzero quantities in the two accounts is suboptimal.

The One-Period Pricing Problem

Let us solve the one-period pricing problem first, starting from the upper bound. We fix a node in the tree and assume we know the derivative's values H_u and H_d in the next period's up-state and down-state, respectively. The derivative's upper bound on the fixed node is derived by finding the minimum V such that the value of the hedging strategy is higher than or equal to the claim's value in both next period's states:

$$\begin{cases} \Delta Su + B_\ell R_\ell - B_b R_b + CR_c \geq H_u \\ \Delta Sd + B_\ell R_\ell - B_b R_b + CR_c \geq H_d \end{cases} \tag{38}$$

Let us set $B := B_\ell - B_b$, which is the cash (positive or negative) held on the initial node.

Minimizing V under (38) is equivalent to minimizing V under

$$\begin{cases} B \geq 0 \\ \Delta Su + BR_\ell + CR_c \geq H_u \\ \Delta Sd + BR_\ell + CR_c \geq H_d \end{cases} \quad \text{and} \quad \begin{cases} B < 0 \\ \Delta Su + BR_b + CR_c \geq H_u \\ \Delta Sd + BR_b + CR_c \geq H_d \end{cases} \tag{39}$$

Let us then assume constant-proportion collateralization: $C := \gamma V$, $0 \leq \gamma \leq 1$. In this case

$$V = \Delta S + B_\ell - B_b + \gamma V \tag{40}$$

so that we can write

$$\Delta S = (1 - \gamma)V - B \tag{41}$$

The two systems (39) thus become

$$\begin{cases} B \geq 0 \\ (1-\gamma)Vu - Bu + BR_\ell + \gamma VR_c \geq H_u \\ (1-\gamma)Vd - Bd + BR_\ell + \gamma VR_c \geq H_d \end{cases} \quad \text{and} \quad \begin{cases} B < 0 \\ (1-\gamma)Vu - Bu + BR_b + \gamma VR_c \geq H_u \\ (1-\gamma)Vd - Bb + BR_b + \gamma VR_c \geq H_d \end{cases} \tag{42}$$

Let us set:

$$R_{\gamma,z} := (1 - \gamma)R_z + \gamma R_c$$

$$p_z := \frac{R_z - d}{u - d}$$

$$V_z := \frac{1}{R_{\gamma,z}}\big[H_u p_z + H_d(1 - p_z)\big]$$

for $z \in \{\ell, b\}$, and denote by $V_{\max}$ the minimum value of V that satisfies (42).[8] We then have the following.

Proposition 4. *The minimum portfolio's value that allows to over-hedge claim H over one period is given by*

[8]The reason for the max subscript is that $V_{\max}$ is the upper bound.

$$V_{\max} = \begin{cases} V_b = \frac{1}{R_{\gamma,b}}\left[H_u p_b + H_d(1-p_b)\right] \text{ if } H_u[(1-\gamma)d+\gamma R_c] > H_d[(1-\gamma)u+\gamma R_c] \\ V_\ell = \frac{1}{R_{\gamma,\ell}}\left[H_u p_\ell + H_d(1-p_\ell)\right] \text{ if } H_u[(1-\gamma)d+\gamma R_c] \leq H_d[(1-\gamma)u+\gamma R_c] \end{cases} \tag{43}$$

In particular, when $H_u[(1-\gamma)d + \gamma R_c] = H_d[(1-\gamma)u + \gamma R_c]$, *then*

$$V_{\max} = \frac{H_u}{(1-\gamma)u + \gamma R_c} = \frac{H_d}{(1-\gamma)d + \gamma R_c}$$

Proof. *Let us set:*

$$A_z := \left\{ (B,V) : \begin{cases} B \geq 0 \\ [(1-\gamma)u + \gamma R_c]V \geq B(u - R_z) + H_u \\ [(1-\gamma)d + \gamma R_c]V \geq B(d - R_z) + H_d \end{cases} \right\}, \quad z \in \{\ell, b\}$$

We want to calculate $V_{\max} := \min\{v_\ell, v_b\}$, *where*

$$v_z := \min_{(B,V)\in A_z} V, \quad z \in \{\ell, b\}$$

Since $(1-\gamma)u + \gamma R_c > (1-\gamma)d + \gamma R_c > 0$, *the two linear programming problems with constraints* A_ℓ *and* A_b *have the following solutions:*

$$v_\ell = \begin{cases} \frac{H_u}{(1-\gamma)u+\gamma R_c} & \text{if } H_u[(1-\gamma)d+\gamma R_c] > H_d[(1-\gamma)u+\gamma R_c] \\ \frac{1}{(1-\gamma)R_\ell+\gamma R_c}\left[H_u \frac{R_\ell - d}{u-d} + H_d \frac{u-R_\ell}{u-d}\right] & \text{if } H_u[(1-\gamma)d+\gamma R_c] \leq H_d[(1-\gamma)u+\gamma R_c] \end{cases} \tag{44}$$

and

$$v_b = \begin{cases} \frac{1}{(1-\gamma)R_b+\gamma R_c}\left[H_u \frac{R_b - d}{u-d} + H_d \frac{u-R_b}{u-d}\right] & \text{if } H_u[(1-\gamma)d+\gamma R_c] > H_d[(1-\gamma)u+\gamma R_c] \\ \frac{H_d}{(1-\gamma)d+\gamma R_c} & \text{if } H_u[(1-\gamma)d+\gamma R_c] \leq H_d[(1-\gamma)u+\gamma R_c] \end{cases} \tag{45}$$

If $H_u[(1-\gamma)d + \gamma R_c] \geq H_d[(1-\gamma)u + \gamma R_c]$, *by straightforward algebra we have:*

$$\begin{aligned} &\frac{1}{(1-\gamma)R_b + \gamma R_c}\left[H_u \frac{R_b - d}{u-d} + H_d \frac{u - R_b}{u-d}\right] \\ &= \frac{u - R_b}{(u-d)[(1-\gamma)R_b + \gamma R_c]}\left[H_d - H_u \frac{(1-\gamma)d + \gamma R_c}{(1-\gamma)u + \gamma R_c}\right] + \frac{H_u}{(1-\gamma)u + \gamma R_c} \\ &\leq \frac{H_u}{(1-\gamma)u + \gamma R_c} \end{aligned}$$

Likewise, if $H_u[(1-\gamma)d + \gamma R_c] \leq H_d[(1-\gamma)u + \gamma R_c]$*, we have:*

$$\frac{1}{(1-\gamma R_\ell) + \gamma R_c}\left[H_u \frac{R_\ell - d}{u-d} + H_d \frac{u - R_\ell}{u-d}\right] \leq \frac{H_d}{(1-\gamma)d + \gamma R_c}$$

Therefore, $V_{\max} = \min\{v_\ell, v_b\}$ *is given by:*

$$V_{\max} = \begin{cases} \frac{1}{(1-\gamma)R_b + \gamma R_c}\left[H_u \frac{R_b - d}{u-d} + H_d \frac{u-R_b}{u-d}\right] \text{ if } H_u[(1-\gamma)d + \gamma R_c] > H_d[(1-\gamma)u + \gamma R_c] \\ \frac{1}{(1-\gamma)R_\ell + \gamma R_c}\left[H_u \frac{R_\ell - d}{u-d} + H_d \frac{u-R_\ell}{u-d}\right] \text{ if } H_u[(1-\gamma)d + \gamma R_c] \leq H_d[(1-\gamma)u + \gamma R_c] \end{cases}$$

In particular, when $H_u[(1-\gamma)d + \gamma R_c] = H_d[(1-\gamma)u + \gamma R_c]$*, then*

$$\begin{aligned} V_{\max} &= \frac{H_u \frac{R_b - d}{u-d} + H_d \frac{u-R_b}{u-d}}{(1-\gamma)R_b + \gamma R_c} = \frac{H_u}{(1-\gamma)u + \gamma R_c} = \frac{H_u \frac{R_l - d}{u-d} + H_d \frac{u-R_\ell}{u-d}}{(1-\gamma)R_\ell + \gamma R_c} \\ &= \frac{H_d}{(1-\gamma)d + \gamma R_c} \end{aligned}$$

The claim value $V_{\max}$ is thus equal to a discounted expectation of next period's payoff. If the first condition in (43) is satisfied, then the expectation is calculated using the probability p_b, and discounted using $R_{\gamma,b}$, which is a weighted average of the collateral and borrowing returns. Conversely, if the second condition is satisfied, then quantities with subscript ℓ must be used.

Corollary 2. *The minimum portfolio's value that allows to over-hedge claim* H *over one period can be characterized as*

$$V_{\max} = \frac{1}{R_{\gamma,x^*}}\left[H_u p_{x^*} + H_d(1 - p_{x^*})\right] \tag{46}$$

where

$$x^* := \begin{cases} \ell & \text{if } B^* \geq 0 \\ b & \text{if } B^* < 0 \end{cases}$$

Proof. The optimal values for the number of shares Δ *and excess cash* B *are*

$$\begin{aligned} \Delta^* &:= \frac{H_u - H_d}{S(u-d)} \\ B^* &:= (1-\gamma)V_{\max} - \Delta^* S \end{aligned}$$

Straightforward algebra leads to:

$$B^* = \frac{-H_u[(1-\gamma)d + \gamma R_c] + H_d[(1-\gamma)u + \gamma R_c]}{(u-d)[(1-\gamma)R^* + \gamma R_c]}$$

where

$$R^* := \begin{cases} R_\ell & \text{if } H_d[(1-\gamma)u + \gamma R_c] \geq H_u[(1-\gamma)d + \gamma R_c] \\ R_b & \text{if } H_d[(1-\gamma)u + \gamma R_c] < H_u[(1-\gamma)d + \gamma R_c] \end{cases}$$

Noting that

$$H_u[(1-\gamma)d + \gamma R_c] > H_d[(1-\gamma)u + \gamma R_c] \Leftrightarrow B^* < 0,$$

(43) can then be rewritten as

$$V_{\max} = \begin{cases} V_b & \text{if } B^* < 0 \\ V_\ell & \text{if } B^* \geq 0 \end{cases}$$

which is equivalent to (46).

Corollary 3. *The minimum portfolio's value that allows to over-hedge claim H over one period is given by*

$$V_{\max} = \max\left\{ \frac{1}{R_{\gamma,b}}\big[H_u p_b + H_d(1-p_b)\big], \frac{1}{R_{\gamma,\ell}}\big[H_u p_\ell + H_d(1-p_\ell)\big] \right\} = \max\{V_\ell, V_b\} \tag{47}$$

Proof. The proof directly follows from noting that the function

$$R \longmapsto f(R) := \frac{1}{(1-\gamma)R + \gamma R_c}\left[H_u \frac{R-d}{u-d} + H_d \frac{u-R}{u-d}\right]$$

is increasing if and only if $H_u[(1-\gamma)d + \gamma R_c] \geq H_d[(1-\gamma)u + \gamma R_c]$. *This is because the first derivative of f with respect to R is equal to:*

$$f'(R) := \frac{H_u[(1-\gamma)d + \gamma R_c] - H_d[(1-\gamma)u + \gamma R_c}{(u-d)\big[(1-\gamma)R + \gamma R_c\big]^2}$$

The Dual Pricing Problem

Let us now consider the "short-position" problem:

$$\text{minimize } V = \Delta S + B_\ell - B_b + C$$

$$\text{subject to } \begin{cases} B_\ell, B_b \geq 0 \\ \Delta S u + B_\ell R_\ell - B_b R_b + C R_c \geq -H_u \\ \Delta S d + B_\ell R_\ell - B_b R_b + C R_c \geq -H_d \end{cases} \tag{48}$$

where $C = \gamma V$, and let us denote its solution by $V_{\min}$.

Setting again $B = B_\ell - B_b$, and using arguments similar to those in the proof of Proposition 4, we get:

$$V_{\min} = \begin{cases} -\frac{1}{R_{\gamma,b}}\big[H_u p_b + H_d(1-p_b)\big] & \text{if } H_u[(1-\gamma)d+\gamma R_c] < H_d[(1-\gamma)u+\gamma R_c] \\ -\frac{1}{R_{\gamma,\ell}}\big[H_u p_\ell + H_d(1-p_\ell)\big] & \text{if } H_u[(1-\gamma)d+\gamma R_c] \geq H_d[(1-\gamma)u+\gamma R_c] \end{cases} \tag{49}$$

which, equivalently to (47), can be summarized as follows:

$$\begin{aligned} V_{\min} &= -\min\left\{\frac{1}{R_{\gamma,b}}\big[H_u p_b + H_d(1-p_b)\big], \frac{1}{R_{\gamma,\ell}}\big[H_u p_\ell + H_d(1-p_\ell)\big]\right\} \\ &= -\min\{V_\ell, V_b\} \end{aligned} \tag{50}$$

The Claim's No-Arbitrage Price Bounds

An arbitrage-free claim price π must satisfy:

$$-V_{\min} \leq \pi \leq V_{\max} \tag{51}$$

Therefore, every value comprised between the negative of $V_{\min}$ and $V_{\max}$ can be regarded as an arbitrage-free price for the claim with payoff H_u (resp. H_d) in the next period's up-state (resp. down-state). This is summarized in the following.

Proposition 5. *The one-period no-arbitrage price π of the claim with payoff (H_d, H_u) must lie in a no-arbitrage band. Precisely, if $H_u[(1-\gamma)d + \gamma R_c] \geq H_d[(1-\gamma)u + \gamma R_c]$, then*

$$\frac{1}{R_{\gamma,\ell}}\big[H_u p_\ell + H_d(1-p_\ell)\big] \leq \pi \leq \frac{1}{R_{\gamma,b}}\big[H_u p_b + H_d(1-p_b)\big] \tag{52}$$

Conversely, if $H_u[(1-\gamma)d + \gamma R_c] < H_d[(1-\gamma)u + \gamma R_c]$, then

$$\frac{1}{R_{\gamma,b}}\big[H_u p_b + H_d(1-p_b)\big] \leq \pi \leq \frac{1}{R_{\gamma,\ell}}\big[H_u p_\ell + H_d(1-p_\ell)\big] \tag{53}$$

In general,

$$\min\{V_\ell, V_b\} \leq \pi \leq \max\{V_\ell, V_b\} \tag{54}$$

Proof. The proof directly follows from (43), (47), (49), (50), and (51).

We notice that the prices of a long call and a short put satisfy (52). The prices of a short call and a long put satisfy (53). To show this, let us consider the long call case as an example. Let us set $H_u = (Su - K)^+$ and $H_d = (Sd - K)^+$, where K is a given strike. We need to prove that

$$Q := (Su-K)^+[(1-\gamma)d + \gamma R_c] - (Sd-K)^+[(1-\gamma)u + \gamma R_c] \geq 0 \tag{55}$$

If $K \geq Su$, then $H_u = H_d = 0$, and hence (55) is trivially satisfied. If $S_d \leq K < S_u$, then $Q = (S_u - K)[(1-\gamma)d + \gamma R_c]$, which is strictly positive. Finally, if $K < S_d$, then

$$\begin{aligned} Q &= (Su - K)[(1-\gamma)d + \gamma R_c] - (Sd - K)[(1-\gamma)u + \gamma R_c] \\ &= (u - d)[(1-\gamma)K + S\gamma R_c] \geq 0 \end{aligned}$$

which proves the result.

The Multi-Period Problem

By dynamic programming, the multi-period problem can be solved by proceeding backwards in the tree. The resulting algorithm is a slight generalization of the CRR one.

Let us denote by $V(t_k, s_j)$ the value of the upper (resp. lower) bound at time t_k on the node where the asset price is $s_j = S_0 u^j d^{k-j}$, $k = 1, \ldots, N$, and $j = 0, \ldots, k$. We start by setting $V(t_N, s_j) = h(s_j)$ on the nodes at final time t_N. Then, we proceed backwards from time t_{N-1} to time 0, and calculate on each node (t_k, s_j) the corresponding value $V(t_k, s_j)$:

$$V(t_k, s_j) = \max_{z \in \{\ell, b\}} \left\{ \frac{p_z V(t_{k+1}, s_j u) + (1 - p_z) V(t_{k+1}, s_j d)}{(1-\gamma)R_z + \gamma R_c} \right\} \tag{56}$$

where $k = 1, \ldots, N$ and $j = 0, \ldots, k$, and where "max" must be replaced by "min" when calculating the lower bound.

Finally, the derivative's upper and lower price bounds at time 0 are given by the corresponding values of $V(0, S_0)$.

Going to the Limit

Recalling the CRR convergence result, let us set

$$\begin{aligned} \delta &:= \frac{T}{N} \\ u &:= e^{\upsilon\sqrt{\delta}}, \\ d &:= e^{-\sigma\sqrt{\delta}}, \\ R_z &:= e^{r_z \delta}, \ z \in \{c, \ell, b\} \end{aligned}$$

where σ is the positive constant of Section 6, and rates r_z are nonnegative.

To prove convergence, we focus on the upper bound. The procedure for the lower bound is analogous and omitted for brevity. Taking the limit of (56) for $\delta \to 0$, we get:

$$\frac{\partial V(t)}{\partial t} + \max_{z \in \{\ell, b\}} \left\{ \frac{1}{2}\sigma^2 S^2 \frac{\partial^2 V(t)}{\partial S^2} + r_z S \frac{\partial V(t)}{\partial S} - ((1-\gamma) r_z + \gamma r_c)\, V(t) \right\} = 0$$

which coincides with the first HJB equation (29).

Convergence in case of a regular payoff functions h follows because the binomial model is a Markov chain approximation of a controlled GBM, and because of local consistency. We refer to [16] for details.

Remark 4. It is interesting to notice that condition

$$H_u[(1-\gamma)d + \gamma R_c] \geq H_d[(1-\gamma)u + \gamma R_c] \tag{57}$$

becomes (30) in limit for $\delta \to 0$. To prove this, let us set

$$H_u = h(xu) = h\big(xe^{\sigma\sqrt{\delta}}\big)$$
$$H_d = h(xd) = h\big(xe^{-\sigma\sqrt{\delta}}\big)$$

for some $x > 0$. Then (57) is equivalent to

$$h\big(xe^{\sigma\sqrt{\delta}}\big)\big[(1-\gamma)e^{-\sigma\sqrt{\delta}} + \gamma e^{r_c\delta}\big] \geq h\big(xe^{-\sigma\sqrt{\delta}}\big)\big[(1-\gamma)e^{\sigma\sqrt{\delta}} + \gamma e^{r_c\delta}\big]$$

A series expansion gives:

$$\big[h(x) + \sigma x h'(x)\sqrt{\delta} + O(\delta)\big]\big[(1-\gamma)(1-\sigma\sqrt{\delta} + \gamma + O(\delta)\big]$$
$$\geq \big[h(x) - \sigma x h'(x)\sqrt{\delta} + O(\delta)\big]\big[(1-\gamma)(1+\sigma\sqrt{\delta} + \gamma + O(\delta)\big]$$

Rearranging terms, we get:

$$2\sigma x h'(x)\sqrt{\delta} \geq 2(1-\gamma)\sigma h(x)\sqrt{\delta} + O(\delta)$$

or, equivalently,

$$xh'(x) \geq (1-\gamma)h(x) + O(\sqrt{\delta})$$

References

1. Antonov, A., Bianchetti, M., Mihai, I.: Funding value adjustment for general financial instruments: theory and practice (2013). Available online
2. Avellaneda, M., Levy, A., Paras, A.: Pricing and hedging derivative securities in markets with uncertain volatilities. Appl. Math. Financ. **2**, 73–88 (1995)

3. Beneš, V.: Girsanov functionals and optimal bang-bang laws for final value stochastic control. Stoch. Process. Appl. **2**, 127–140 (1974)
4. Bergman, Y.Z.: Option pricing with differential interest rates. Rev. Financ. Stud. **8(2)**, 475–500 (1995)
5. Bielecki, T.R., Rutkowski, M.: Valuation and hedging of OTC contracts with funding costs. Collateralization and Counterparty Credit Risk: Part 1. Working paper. Illinois Institute of Technology, Chicago (2013)
6. Burgard, C., Kjaer, M.: Partial differential equation representations of derivatives with counterparty risk and funding costs. J. Credit Risk **7(3)**, 1–19 (2011)
7. Castagna, A.: Pricing of derivatives contracts under collateral agreements: liquidity and funding value adjustments (2011). http://papers.ssrn.com/sol3/papers.cfm?abstract_id=1974479
8. Cox, J.C., Ross, S., Rubinstein, M.: Option pricing: a simplified approach. J. Financ. Econ. **7**, 229–263 (1979)
9. Crépey, S.: Bilateral counterparty risk under funding constraints - part I: pricing. Mathematical Finance, **25(1)**, pp. 1–22 (2015). doi:10.1111/mafi.12004
10. Crépey, S.: Bilateral counterparty risk under funding constraints - part II: CVA. Mathematical Finance, **25(1)**, pp. 23–50 (2015) doi:10.1111/mafi.12005
11. El Karoui, N., Peng, S., Quenez, M.-C.: Backward stochastic differential equations in finance. Math. Financ. **7(1)**, 1–71 (1997)
12. Forsyth, P.A., Labahn, G.: Numerical methods for controlled Hamilton-Jacobi-Bellman PDEs in finance. J. Comput. Financ. 1–43 (2007)
13. Fujii, M., Takahashi, A.: Derivative pricing under asymmetric and imperfect collateralization and CVA. Quant. Financ. **13(5)**, 749–768 (2013)
14. Jouini, E., Kallal, H.: Arbitrage in securities markets with short-sales constraints. Math. Financ. **5(3)**, 197–232 (1995)
15. Korn, R.: Contingent claim valuation in a market with different interest rates. Math. Meth. Oper. Res. **42**, 255–274 (1995).
16. Kushner, H.J., Dupuis, P.: Numerical Methods for Stochastic Control Problems in Continuous Time. Springer, New York (2001)
17. Laurent, J.-P., Amzelek, P., Bonnaud, J.: An overview of the valuation of collateralized derivative contracts (2012). http://events.em-lyon.com/AFFI/Papers/10.pdf
18. Leippold, M., Su, L.: Collateral smile (2011). http://papers.ssrn.com/sol3/papers.cfm?abstract_id=1956449
19. Mercurio, F.: Differential rates, differential prices. Risk 100–105 (2014)
20. Pallavicini, A., Perini, D., Brigo, D.: Funding, collateral and hedging: uncovering the mechanics and the subtleties of funding valuation adjustments (2012). http://arxiv.org/pdf/1210.3811.pdf
21. Piterbarg, V.: Funding beyond discounting: collateral agreements and derivatives pricing. Risk **2**, 97–102 (2010)
22. Rendleman, R.J. Jr., Bartter, B.J.: Two-state option pricing. J. Financ. **24**, 1093–1110 (1979)

Index

J.A. Londoño et al. (eds.), *Actuarial Sciences and Quantitative Finance*, Springer Proceedings in Mathematics & Statistics 135, DOI 10.1007/978-3-319-18239-1

MIX
Papier aus verantwortungsvollen Quellen
Paper from responsible sources
FSC® C105338

If you have any concerns about our products,
you can contact us on
ProductSafety@springernature.com

In case Publisher is established outside the EU,
the EU authorized representative is:
Springer Nature Customer Service Center GmbH
Europaplatz 3, 69115 Heidelberg, Germany

Printed by Libri Plureos GmbH
in Hamburg, Germany